RADIATION ONCOLOGY CONCEPTS

Order this book online at www.trafford.com
or email orders@trafford.com

Most Trafford titles are also available at major online book retailers.

Note for Librarians: A cataloguing record for this book is available from Library and Archives Canada at www.collectionscanada.ca/amicus/index-e.html

Printed in Victoria, BC, Canada.

ISBN: 978-1-4269-1653-3

Our mission is to efficiently provide the world's finest, most comprehensive book publishing service, enabling every author to experience success. To find out how to publish your book, your way, and have it available worldwide, visit us online at www.trafford.com

Trafford rev. 8/15/2009

www.trafford.com

North America & international
toll-free: 1 888 232 4444 (USA & Canada)
phone: 250 383 6864 • fax: 812 355 4082

Authors

Brian Purdie, M.D.

Emory University School of Medicine
Hospital Medicine
Instructor

Anna Harris, M.D.

University of Texas Health Science Center at San Antonio
Radiation Oncology
Resident

Jonathan J. Beitler, M.D., MBA, F.A.C.R.

Emory University School of Medicine
Professor of Radiation Oncology
Professor of Otolaryngology
Professor of Hematology and Medical Oncology

Table of Contents

Introduction

The Radiation Oncology Concepts is intended to be a primer for medical students, non-radiation oncology residents, as well as medical oncology fellows who will be spending time in the clinical setting of a radiation oncology (RadOnc) office. We hope to provide a useful overview to those unfamiliar with RadOnc, and more specifically to demonstrate how the Radiation Oncologist works with surgical oncologists, medical oncologists, physicists, dosimetricists, therapists, nurses, and others. Because RadOnc is not a primary care field, and many medical students are not exposed to the specialty during medical school, we feel this book will address that niche.

The surgeon's role in cancer management may range from biopsying lesions, debulking tumor, or complete resection of the tumor mass. Medical oncologists try to cure or manage cancer principally with the use of systemic medications. On the other hand, many medical professionals have no exposure to the processes and purposes of radiation oncology. In the interest of patient care, it is important for other physicians to understand the function of the radiation oncologist when cancer patients are initially diagnosed, undergoing concomitant chemotherapy, or in need of palliative pain control.

For those considering entering radiation oncology, entry can only be described as competitive. In addition to providing a fascinating opportunity to help patients at a crucial point in their lives, a high reimbursement rate, and an attractive lifestyle, this specialty allows one to grow with the changing technology. Because of the very few RadOnc residency positions at accredited medical institutions, radiation oncology program directors have their pick of well-qualified candidates and are looking for dedicated people who can help build their programs. Characteristics of a good applicant include someone who gets along well with others, bringing with them laboratory skills, advanced degrees in basic science, and academic potential. This can be demonstrated with journal publications or participation in clinical research.

One of the strengths of RadOnc is that few patients with cancer come directly to the radiation oncologist's doorstep. Radiation oncology is supplied by patients almost exclusively through referrals. This implies that radiation oncologists have demonstrated to other medical specialties the scientific data supporting the use of radiation either: 1) in addition to, or 2) instead of other modalities. Decisions are often made in consultation with other physicians, the patient, and patient's family. Therefore, the radiation oncologist must communicate what is known about the patient's condition and be candid about what is not known.

Both written and oral boards are part of the four certifications of the specialty. The written boards consist of three separate tests: clinical radiation oncology, physics, and radiobiology. The mastery of all three subjects is a prerequisite to the oral examination for complete board certification. Currently, new applicants can obtain board certification that is good for ten years. Radiation oncology is technology and science driven, requiring considerable dedication to staying current. Nonetheless, RadOnc residents are some of the most content among the spectrum of medical residencies, and practicing radiation oncologists remain among the most satisfied with their career choice.

Because of the complexity and ongoing research into various treatments for individual types of cancer, we will leave them aside. The focus of this book will be to build on the fundamentals which will help you understand some of the processes within the radiation oncology departments.

The Application Process

Advisors, Grades, and Step 1

Exposure to RadOnc may be limited at your institution as a medical student. While many RadOnc departments are beginning to expand, it still remains a small field. Residency programs can be found at larger facilities that serve as a tertiary referral center or that of a cancer center. As a first and second year medical student you may try to find an advisor in the RadOnc or medical oncology department. Given the competitiveness of the RadOnc match, it is best to get a head start on preparing for the USMLE Step 1. Whereas no one wants to get a bad score on Step 1 or 2, it is well noted that Step 1 carries an emphasis on basic science that is weighed more heavily on your application. The average step 1 score for a resident matching in RadOnc varies by year, but remains typically in the 230-240 range. The broader spectrum is generally from 220 to 260 or above, but those with other outstanding qualities on their resume may have lower scores. Grade point average is typically 3.5-4.0 for those matching into RadOnc. Be aware that your clinical rotations as a third and fourth year medical student are weighted more heavily than your didactic classroom studies of first and second year of medical school. It is important to not only shine on your clinical rotations but also to perform well on your end of rotation tests. Be aware: an 'A' from your grading faculty member plus a 'C' on your rotation test will average you out to a 'B' every time.

Research, Rotations, and Letters of Recommendation

Research is also an important part in gaining an edge in the match. The best opportunity to get involved in research may be during your summer weeks between first and second year of medical school. Many medical schools offer positions to assist in basic research that will stand out on your resume. Approaching RadOnc residents and asking to assist on research projects may also be helpful. The reception you receive as an eager medical student interested in RadOnc may vary from discouraging to welcoming depending on the department at your institution. Despite this, you will need to do a clinical rotation with your home RadOnc department (if you have one) as early as possible, either during your third year or the first month of your fourth year. Rotating early is essential to collect letters of recommendation for the match. Generally you want letters of recommendation from the faculty who know you best personally and know of your work with the department. Name recognition among your letter writers in such a small subspecialty as RadOnc is advantageous, and a letter from the chairman of your department may be beneficial. If you do not have a home department, you may need to speak with your dean to receive permission to do an away rotation. This provides students with a limited amount of time in which to integrate themselves into a department and get to know everyone, so consider a month at one or two away institutions. Also, use these opportunities to work with residents and faculty on potential research projects.

AΩA

The addition of Alpha Omega Alpha (AΩA) medical honor society on your application is not a necessity, but it is certainly helpful. The first AΩA class is typically designated during the third year of medical school. There is also a smaller portion of students who make it their fourth year. Even those in internship may apply for AΩA, so don't give up.

ERAS

All RadOnc programs currently use the Electronic Residency Application Service (ERAS) to process your application, and the RadOnc match, although competitive, is not an early match. The RadOnc residency is a 5 year program that is broken into a transitional or medicine year plus four years within the structure of the RadOnc program. Do your research when applying to RadOnc programs, as some offer or may even mandate that you complete your first year of training at their medical facility. You also need to apply to many programs given the competitiveness of RadOnc. Whereas couples matching is certainly not impossible, it does pose much difficulty when a spouse is also applying to a moderately or very competitive specialty (i.e. dermatology, plastic surgery).

There are a number of Ph.D. candidates applying to RadOnc. An inclination to select residents with a prior concentration in physics and biology is certainly present, but this should not discourage applicants, nor should anyone without a background in these subjects seek out an additional degree.

Residency

The PGY-1 year entails clinical training in internal medicine, pediatrics, surgery, or a combined transitional year. Some programs may detail the type of PGY-1 year required to gain acceptance to their program, and others mandate you complete the PGY-1 year at their affiliated institution.

As a PGY-2, a resident is expected to apply basic knowledge to perform an appropriate history and physical, present the case summarizing important factors and findings, recommend staging tests, and discuss general principles of treatment and the role of radiation with approximate treatment results. The resident is also expected to communicate this plan to the patient and family, answering questions regarding treatment outcome and toxicity.

The PGY-3 year requires a resident be capable of planning and defending a staging workup, evaluating results of clinical investigations, proposing and defending a management plan possibly including surgery, radiation, and chemotherapy for common tumors, as well as communicating the plan to the patient & family.

PGY-4 and PGY-5 residents should be able to formulate a management plan, discuss controversies of management citing key literature in the area with level of evidence. The PGY-5 resident should be able to function nearly independent of an attending.

Salary and Positions

Nearly three quarters of physicians in radiation oncology are in private practice. The remaining are at academic institutions, but nearly all are hospital based and associated with a cancer center. A good relationship with medical oncology and surgery, particularly in a setting with a tumor board conference, assists in developing the best treatment for patients.

Different studies from 2003 show radiation oncologists making a wide range of salaries from $250,000 first year after residency to upwards of $700,000, working an average of 55 to 60 hours a week.

Malpractice premiums for the specialty are low, and there are few oncologic emergencies requiring immediate radiation treatment. Therefore, most physicians can take home call. Much of a radiation oncologist's spare time is spent keeping up with the latest medical literature. The newest technology is always advancing, and cancer treatment plans are continually challenged with new clinical trials.

History

Radiotherapy would not have come about were it not for the discovery of x-rays by Wilhelm Roentgen in 1895, an achievement for which he received the Nobel Prize in physics in 1901. Radium, the radioactive element discovered by Marie and Pierre Curie, and uranium, discovered by Henri Becquerel, also played a role in early radiotherapy treatments. Radiotherapy was initially used on skin cancer patients with some success. Patients were first given large, single doses of radiation, a treatment known as the massive dose technique. The side effects of such a dose caused complications, and there was a very high level of tumor recurrence.

The idea of fractionation would later be proven to be a more effective method of radiation delivery. Claude Regaud, a Frenchman, used various doses of radiation on the genitalia of rams. The process of spermatogenesis was thought to be the closest imitator of cancer cells at work. By giving large single doses of radiation, the rams experienced erythema and skin desquamation. However, when the doses where given in small quantities over a longer period of time, the skin effects were much less noticeable, and the rams were still found to be sterile at the end of the treatment course. This indicated that the end result of halting spermatogenesis could be attained over an extended timeframe by administering fractions of radiation with better tolerance.

It was actually a medical student named Emil Grubbe from the Hahnemann Medical College of Chicago that is known as the world's first radiation oncologist. After working with x-rays and suffering radiation dermatitis of the hands, Grubbe put x-rays to work on breast cancer patients. Achieving adequate results, he founded the first radiation oncology facility in Chicago before he even graduated medical school.

SECTION ONE:

RADIOBIOLOGY

"As pharmacology is to the internist so is radiation biology to the radiotherapist[1]" said a classic textbook by Gilbert Fletcher. Classic radiobiology was simplified into the four R's which we will go into momentarily. The big picture is to recognize that there is a major discrepancy between tumor cells and normal cells. If we use physics to treat the cancer and spare the normal tissue, then radiobiology helps us to expand the therapeutic index by doing more to the tumor and less to the normal cells.

1. Fletcher GH: Textbook of Radiotherapy, (ed Third). Philadelphia, Lea & Febiger, 1980, pp 959

Normal Tissue versus Tumor

One of the underlying concepts of radiation is that cancer cells are intrinsically more sensitive to radiation than normal tissue. The theory behind this may be as simple as the cancer cells are more likely to be in a radiosensitive phase of the cell cycle, whereas most normal cells are quiescent in the G_0 phase. Other mechanisms to consider are the cancer cell's decreased ability to self-repair.

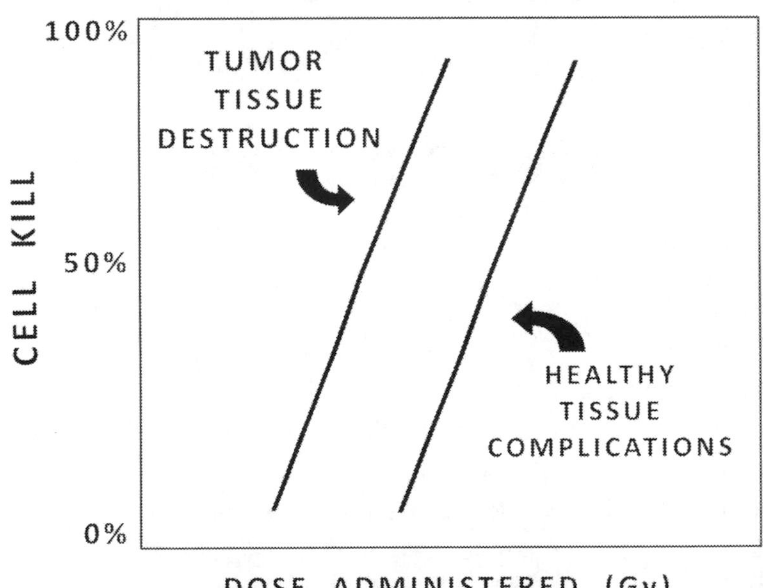

The 4 R's of Radiobiology

After each radiation treatment the dynamic of the tumor environment changes greatly. There are four mechanisms that occur, and a greater understanding of them explains why fractionation of therapy is so important.

1. Repair
2. Repopulation
3. Redistribution
4. Reoxygenation

Repair refers to the concept of DNA repair after ionization and free radical damage caused by radiation. Some damage may not be amenable to repair and repeated damage may have a lethal effect after several doses of radiation. Ideally, a healthy cell has the ability to repair itself in time before the next fraction of radiation is delivered, whereas a tumor cell may not have this capability.

Repopulation (a.k.a. regeneration) is the principle that normal tissues and tumors are capable of increasing their cell replication rate in response to depopulation caused by radiation. A more aggressive tumor may repopulate sooner than a less aggressive malignancy. By this same concept, healthy skin tissue that regenerates very rapidly will also allow for quicker healing. As the cells within the irradiated region die off, there is a new mix of healthy and tumor tissue.

Redistribution pertains to those cells that survive after radiotherapy and are able to proceed through the cell cycle. The cells that are 'caught' in the radiosensitive stage of the cell cycle are more likely to undergo DNA damage and subsequent apoptosis. In between radiotherapy fractions, the surviving cells will be 'redistributed' to various stages of the cell cycle.

Hopefully, more tumor cells will be in a vulnerable state of the cell cycle by the time the next fraction is administered.

Finally, reoxygenation pertains to the restructuring of the tumor's vascular supply during the death and subsequent decrease in tumor cell mass after radiotherapy. Those portions of the tumor which were previously hypoxic may find easier access to vasculature, resulting in increased oxygen levels. Hence, these cells become more radiosensitive, leading to additional tumor cell death. The vascular supply can continue to invade hypoxic regions with each fractionation.

Dose Response Curve

The dose response curve demonstrates the relationship between an organism and an associated stressor and is fundamental to radiobiology. In the context of radiobiology, we are using this relationship to determine how a tumor cell reacts to its stressor, radiation. Different tumor types and treatment regimens may yield widely varying dose-response curves. Note the shoulder region of the curve during the initial stages of radiotreatment.

At first, the cell may be able to repair itself and damage may be minimal. However the "sublethal damage" (described later) may eventually overwhelm the tumor and healthy tissue before a steady cell kill develops.

The two-hit theory of radiation implies that a normal cell can survive one ionization, and until a substantial portion of the normal cells have had at least one insult, there will be limited normal cell killing (this is depicted as the shoulder region). Once this quasi-threshold has been passed and all of the normal cells have had their one hit, imagine that another ionization occurs.

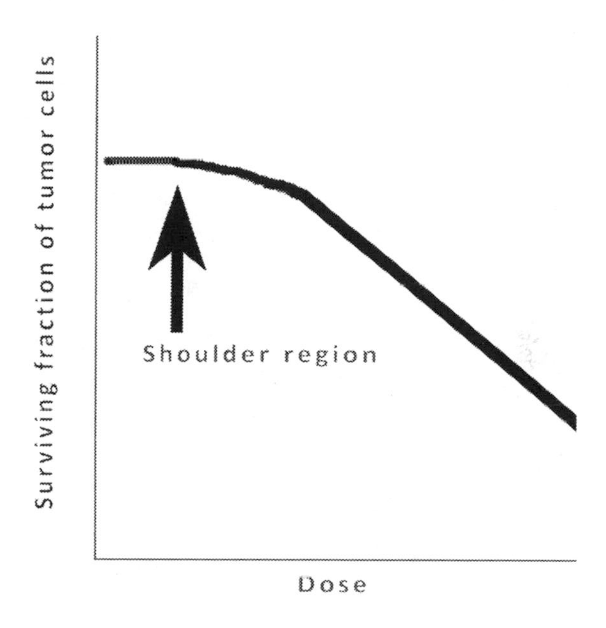

For this model, it represents the second hit, and the threshold has been passed. All additional "hits" cause cell kill. You then reach the steeper part of the curve shown above. The assumption is that cancer cells are less capable in making these self-repairs after a single hit, and they have little or no threshold before they reach the steeper part of the cell survival curve.

Ionization

How does radiation work? The energy is deposited as packets within atoms known as spurs (the smallest), blobs, or short tracks (the largest). Depositing energy within an atom results in ejection of the electron from its orbit, thus causing transformation of that atom or molecule into a free radical. The free radical may be formed directly as when DNA itself is irradiated, or indirectly when a reactive oxygen species (ROS) is formed. The ROS species may then transfer an electron to DNA. It is the generation of oxygen free radicals created by radiotherapy that results in damage to numerous molecules within the cell, not just DNA. The generation of double strand breaks (DSB) to DNA, however, accounts for 70% of radiation-induced cell kill. Just one DSB can lead to cell death if it affects a crucial portion of the DNA.

Radiation produces cell kill via two main mechanisms:

1. Reactive oxygen species
   - H_2O_2
   - $OH\bullet$
   - $HO_2\bullet$
   - $O_2\bar{\bullet}$
   - OCl^-
2. DNA double strand breaks

The cluster hypothesis of free radical production states that those molecules within the localized area of the initially produced free radical will undergo the greatest amount of change given their proximity.

The radiation bystander effect implies that radiation injury may occur in cells adjacent to, but not part of radiation therapy targets. These cells may be affected by indirect energy deposition or production of free radicals that have migrated from irradiated cells.

Linear Energy Transfer

Linear energy transfer (LET) is the concept of energy delivered by radiation, as expressed in units of charge and mass. This unit is measured in kiloelectron volts per micrometer (keV/μm). Electrons have almost no mass, thus exhibiting a low LET, whereas neutrons are relatively dense particles exhibiting a high LET. Cell damage resulting from a high LET is expected to be much greater than that of a low LET. This is illustrated below: notice the higher dose of photons required for cell kill compared with the dose of neutrons.

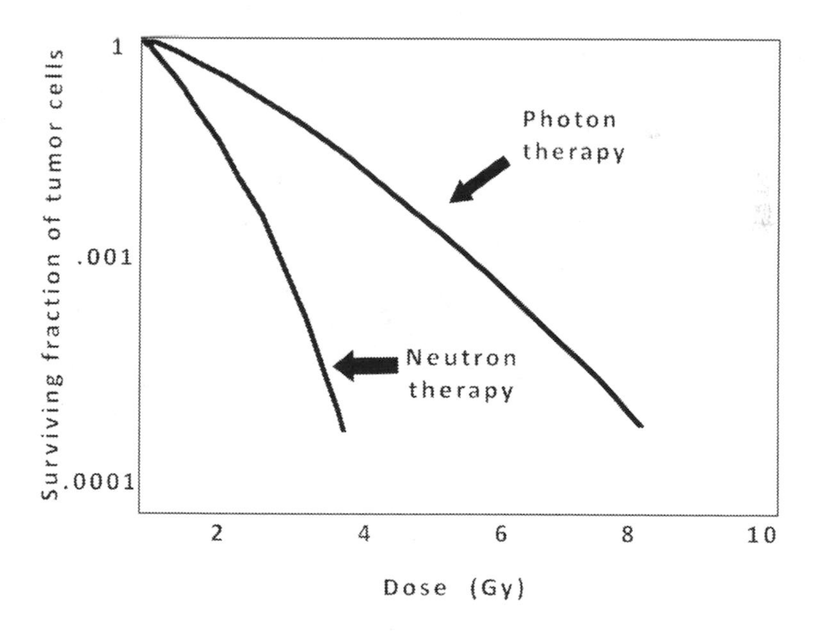

It is important to know the relative biological effectiveness (RBE) of a LET value because it relates the different modalities of radiotherapy to the different levels of damage within a biological material. A higher RBE means that there is a greater amount of biological damage as a result of radiation exposure. For standardization, photons are assigned the base value of 1 for their RBE. Thus, RBE is a ratio of the dose of a photon to that of a higher LET type of radiation. When LET reaches 100 keV/µm there is maximal biological effectiveness, and the RBE value is at its highest. Radiation at this level has the highest likelihood of creating DNA double strand breaks (DSB). Delivering more than this amount results in a waste of energy. The result is an 'overkill effect', as too many ionizing effects will occur within the cell without increasing the probability of DSBs.

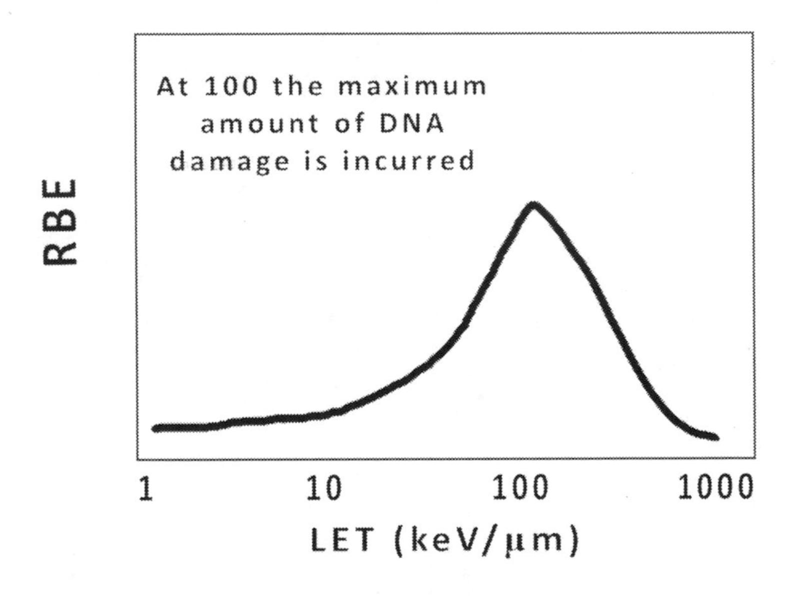

RBE may vary by the type and dose of radiation delivered, as well as the fractionation pattern. For a given dose and time span, a more fractionated regimen will generate a higher RBE.

Apoptosis

Apoptosis is otherwise known as programmed cell death (PCD). Healthy tissue and tumor cells are apt to undergo PCD after irradiation. The cells may first exhibit nuclear condensation and membrane blebbing prior to eventual lysis of the cell wall within 12 hours of entering the PCD pathway.

> The definition of cell death in radiation oncology may differ from that of a biologist. Cell death accounts for those tumor clonogens that have lost the ability to reproduce. This includes the potential for the cells to be induced into senescence (remaining metabolically active without proliferating).

Radiation oncology is faced with the undesirable situation in which tumor cells that survive are those that have resistance to programmed cell death while normal cells retain these pathways. The ability of cells to inhibit the normal pathways of cell death is a key step in the origin of cancer cells.

Many tissues have been experimentally irradiated, and various levels of radiosensitivity have been documented. Certain healthy tissue is susceptible to radiation, causing PCD. The hematopoietic, lymphoid, crypt cells of the small intestine, and salivary glands are easily destroyed by radiation, entering PCD shortly after therapy.

The Cell Cycle

The cell traverses a one-directional cell cycle as shown below:

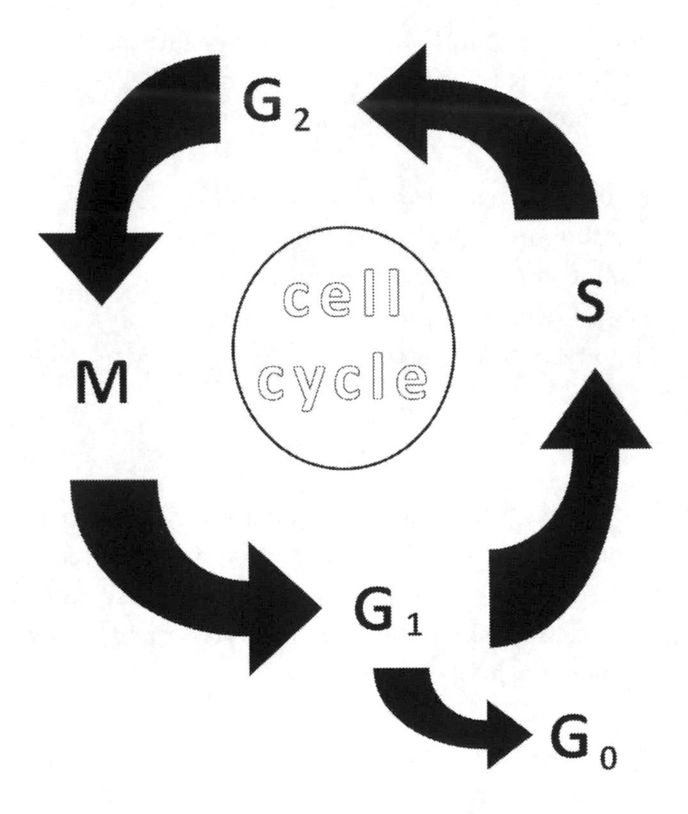

This process is driven by a sequence of protein kinases known as cyclin-dependent kinases (CDK) that are activated to allow the cell to enter each stage. Each stage of the cell cycle has its own set of CDKs that must be activated by phosphorylation before the cell can progress to the next stage. Because proteasomes act as degradation enzymes of proteins within the cell, they destroy each cycle's CDKs after their use to negatively control movement through the cell cycle.

There are 4 basic cycles of reproducing cells: G_1, S, G_2, and M. Resting cells, also known as senescent cells, are those that rarely (if ever) replicate.

Note: Understand that all movement through the cell cycle is one-way without the possibility of shifting back after a phase is completed.

Furthermore, some cells are in such a steady state that they rarely or never replicate, this is known as *senescence*. Other cells that retain the ability to divide but are 'resting' are known to be in the growth arrest, G_0, or *quiescent phase*.

Cells appear most radioresistant in the S and early G_1 phase. Radiosensitivity begins at the end of G_1, with peak sensitivity within the late G_2 and M phase (G_2/M). The increased vulnerability at the time of mitosis can be explained by the high likelihood of subsequent *mitotic catastrophe*. This results in difficulties constructing the spindle apparatus, leading to immediate death or uneven distribution of chromosomal complement to the daughter cells.

Also, the cells in M phase are at their most vulnerable in the telomere phase during which the parent cell membrane divides into its daughter cells, hence, losing contact with the stromal milieu during this phase. This results in a vulnerability to death or senescence that is not counteracted by the normally protective integrin mediated pathways.

In some cases, DNA damage is so significant that the cell is unable to carry out mitosis. Cells that survive radiation generally move through the cell cycle and stall within the S or G_2 phase where DNA repair can take place. In other cases, the new cell is only able to survive for a short time after cell division. Some cells may even replicate after the damage has taken place, eventually succumbing to death. This concept can be further understood after a developing a better understanding of the DNA repair mechanisms below.

Note: The movement of the cell through the cell cycle is under the control of the checkpoint genes.

In one example of cell cycle arrest, the protein CDKN1A acts as an inhibitor to CDKs at the G_1 checkpoint. CDKN1A accumulates in irradiated cells preventing the progression beyond G_1. The TP53 protein product is found in the nucleus of cells that assists in the transcription of CDKN1A and other important products regulating the cell. As a result of much research, TP53 is considered the master regulator of apoptosis and has been shown to be inactivated in many tumor cell lines. Loss of TP53 function in a malignancy is associated with a worse prognosis as these cells are more radioresistant. The significance of TP53 loss within the cell lies with the DNA damage that is never corrected as the cell continues through its life cycle. With further cell division of unrepaired

DNA, there is the creation of more unrepaired DNA damage and tumor heterogeneity develops.

DNA Repair

DNA damage may occur at several levels as a result of irradiation:

1. DNA base damage
2. Deoxyribose sugar damage
3. Single strand DNA breaks (SSB)
4. Double strand DNA breaks (DSB)

The response to DSBs by radiation are two-fold:

1. Stop cell division
2. DNA repair

As a result of toxicity, some cells are beyond repair and enter programmed cell death (PCD) pathways. For tumor cells, this is ideal. For normal tissue, the cells must repair themselves under the stress of the administered radiotherapy.

DNA is the only molecule with its own repair system. There are several pathways of DNA repair to become familiar with.

Base, Nucleotide, and Mismatch Repair

Base excision repair is controlled by a series of repair enzymes. DNA glycosylases, AP endonucleases, and exonucleases cut out damaged bases. The DNA polymerase patches the strand of missing DNA, and DNA ligase seals them back into place. The mechanism of nucleotide excision repair relies on structure specific endonucleases to recognize gross abnormalities in the DNA configuration.

Mismatch repair seeks to correct DNA errors made by DNA polymerase when incorrect base pairing occurs. This requires recognition and degradation of the incorrect strand, followed by repair. Defects in this repair mechanism have been implicated with hereditary nonpolyposis colon cancer (HNPCC).

NHEJ repair of DNA DSBs

When DNA strands break, it is difficult to repair. There may be no template to assist in the repair process. A mechanism known as non-homologous end joining (NHEJ) reattaches DNA that is broken on both strands. Portions of genetic material on the end of each strand pose risk of deletion in this process. It is thought that BRCA1 and BRCA2 gene products interfere with this DNA repair mechanism, leading to breast and ovarian cancer.

The first step in DNA double stand break (DSB) repair is the recognition of damage. Irradiated cells activate production of the RAD proteins. The RAD proteins then migrate to the nucleus to assist with recognizing DNA damage. Loss of the RAD proteins may play a role with BRCA1, TP53, and Ataxia telangiectasia in cancer development.

DSB repair is mediated by a number of genes, notably the XRCC family which codes for various proteins including XRCC5, XRCC6, and DNA-PK. XRCC5 protein binds to the ends of double stranded DNA with XRCC6 and DNA-PK proteins preventing degradation of the end portion of the DNA. This complex works with DNA ligase IV to catalyze the final step in NHEJ repair adjoining the pair of DNA strands.

Wortmannin, a molecule discovered in fungi, has been found to inhibit the activity of DNA-PK. It has been shown to allow for greater radiosensitization of tumor cells as DSBs are not able to be repaired.

> Note on Ataxia telangiectasia:
>
> Cells with ataxia telangiectasia (AT) defects do not recognize the presence of DNA damage within the region of the telomeres. The telomeres in AT cells are seen to exhibit more chromosomal instability than healthy cells. Although patients with AT are thought to be sensitive to radiation, it is merely their inability to repair DNA damage produced by such radiation that is so detrimental.

Sublethal Damage

Recall Regaud's experiments with rams and the clinical results of less skin irritation with fractionation. One explanation of why fractionation works on a cellular level is thought to be due to sublethal damage repair (SLDR), also known as sublethal damage recovery. When radiation treatments are broken into fractions, there is less damage to healthy tissue. These healthy cells are believed to retain their DNA repair mechanisms. Working between fractionated treatments to repair DNA damage allows these cells to preserve their original genetic makeup and retain their healthy existence.

We previously discussed the difference between high and low LET. Due to its lower level of damage, a significant amount of sublethal damage is associated with low LET. This concept is nearly abolished when the practice of high LET is applied.

The survival curve of radiated cells is an exponential one with the exception of the initial “shoulder” region. At very low radiation doses, expected cell kill does not occur at a rate consistent with the remaining slope of the dose-response curve.

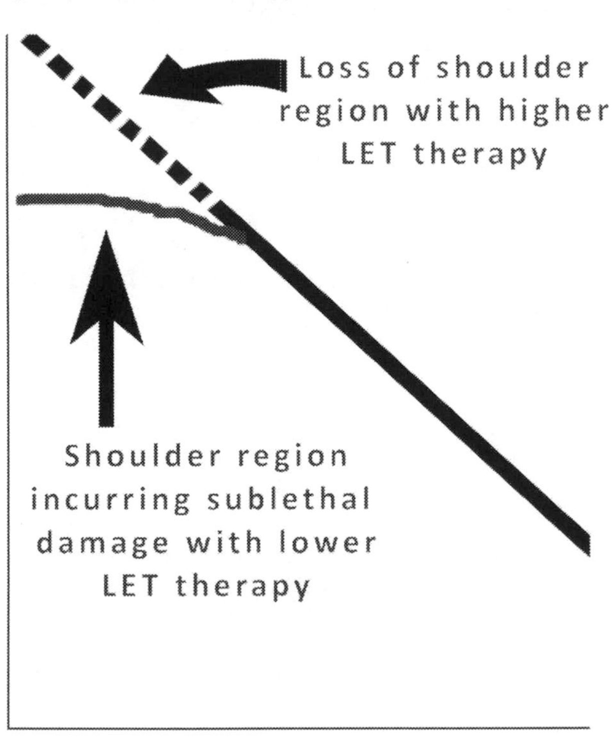

The shoulder portion of the curve results from DNA repair mechanisms still intact within the cell for repairing sublethal damage. At a lower dose, it is presumed that the cell was able to survive a small amount of sublethal damage. After subsequent radiation (i.e., after a higher 'lethal' dose of radiation has been delivered), the cell's sublethal damage accumulates, now resulting in lethal damage. This decreases the number of surviving tumor cells.

Sublethal damage repair can also be affected by the frequency of treatments. If 1.8 gray of radiation is administered, and another 1.8 gray treatment is given immediately thereafter, the effect on cell death is much greater because DNA repair from the first treatment has not yet been completely carried out. A cell that was previously stunned by an initial radiation treatment is now knocked out with a second treatment before it can recuperate.

> Linear Energy Transfer (LET) is a measurement of a radiation's density of ionization along a particle's path. Sublethal damage may be too severe in some cases of higher LET, thus SLDR may not be able to protect cells from dying.

It should be noted that resting cells involved in radiation treatments will also incur damage. These resting cells have the advantage of allowing for more time to undergo DNA damage repair, a theory known as potentially lethal damage repair (PLDR).

Healthy cells would be expected to repair DNA damage faster than tumor cells. If the radiation treatments are too widely spaced, tumor cells themselves may be able repair their DNA damage between fractions, thus allowing for radioresistance of the malignancy.

Tumor Heterogeneity

The idea that a malignancy begins with only one cell transformation may be true, however, as the tumor proliferates, the genetic material begins to vary greatly. The genetic abnormalities initiating proliferation become greater and more complex as the malignancy matures. This idea is referred to as tissue heterogeneity. Many different subpopulations of cancer cells can make up a tumor, and each may vary in their resistance to chemotherapy and radiotherapy. More heterogeneous tumors tend to have a shallower dose-response curve, as they may contain a greater number of subpopulations of tumor cells. This mix of cells may contain cells that vary widely in their radiosensitivity.

Cell Replication and Sensitivity

There is a correlation between how many times a cell replicates and its sensitivity to radiation. Cells undergoing more frequent replication have less time to repair DNA damage.

Vegetative intermitotic cells (VIM)
These are the most radiosensitive cells because they are mostly undifferentiated stem cells that are continuously undergoing replication. Cells such as those found in the blood, bone marrow, testis, gastrointestinal tract, and basal layer of the skin are good examples.

Differentiating intermitotic cells (DIM)
Though not completely undifferentiated, these cells have not yet developed. As a result, they are less radiosensitive than VIMs. The spermatocytes are DIMs because they have a small number of cell divisions before they can become mature sperm.
Reverting postmitotic cells (RPM)
RPMs do not regularly divide and are more radioresistant. Cells like the hepatocytes have the capability of dividing, but rarely do so.
Fixed postmitotic cells (FPM)
As terminally differentiated cells, FPMs do not divide. They are the most radioresistant to radiotherapy. Neurons are considered nonreplicating FPMs.

Cells rapidly dividing and susceptible to radiation damage are also the first to recover. This will allow them to be re-irradiated at an earlier date. Lung and nervous tissue take longer to recover and may only be re-irradiated at a lower dose than the initial treatment given. Bladder tissue damage is often permanent, and re-irradiation may be performed at a significantly reduced level and sometimes discouraged altogether.

As cells survive one treatment course of radiotherapy, their response to a repeated regimen may vary depending on the characteristics of the tissue type:

1. Cells may maintain the same response they had prior to the first course of treatment.
2. Cells may become better able to tolerate re-irradiation.
3. Cells may become easier to kill because they remain with DNA damage from the initial treatment.

Isoeffect Curves

The concept of radiating various tissues at different doses and with different fractionation schedules has been researched extensively. This has been done to look at optimizing the amount of tumor cell kill while at the same time, accounting for harmful side effects on normal tissue. The results can be plotted on a graph, and each tissue has a specific isoeffect curve. The isoeffect curve plots the total dose necessary to produce a certain endpoint in tissue being radiated versus the dose per fraction.

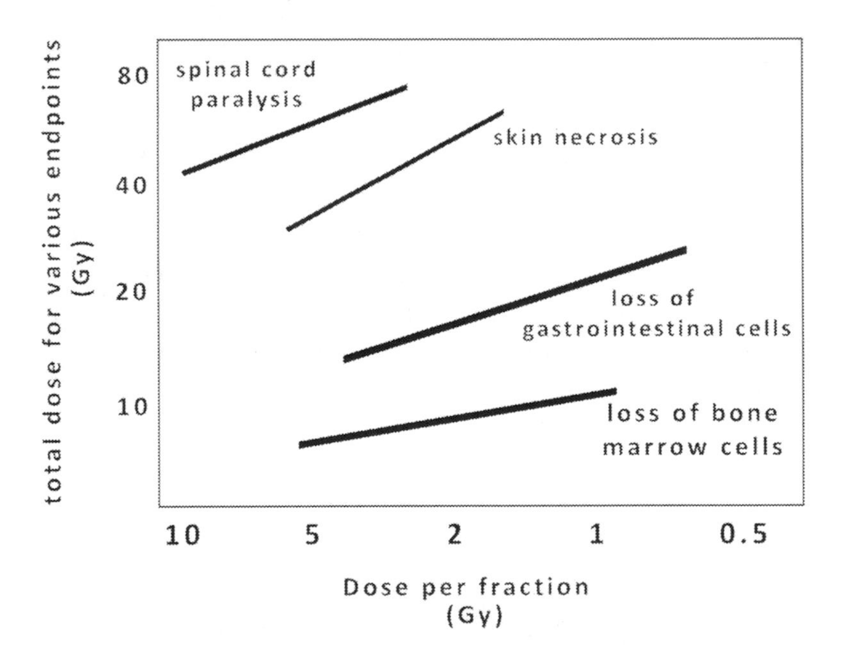

By utilizing the isoeffect curve, one may be able to plot whether or not side effects can be avoided. By keeping the same total dose, an isoeffect curve can help determine if a certain endpoint may be avoided by decreasing the amount of the dose used during fractionation. In fact, one can determine whether or not a larger total dose can be given in smaller increments without achieving any adverse outcomes.

Notice the diagram above for those tissues exhibiting steeper isoeffect curves, like that of skin necrosis. The adverse side effect of skin necrosis may occur with as little as 25 Gy total dose or 75 Gy total dose. The isoeffect curve demonstrates that we may give a total of 75 Gy at a dose of 1.5 Gy per fraction and that the same adverse effect of skin necrosis will occur when we deliver a total of 25 Gy at a dose of 6 Gy per fraction. This also follows the logic of the previous work initially performed by Regaud. Therefore, when treating a tumor, we would prefer to be able to deliver 75 Gy total dose, if indicated.

E/α

The biological effective dose (BED) is a way to quantify treatment expectations among different fractionation schemes and/or different cumulative doses. The BED is also known as 'E/α' from its derivation of the original equation (see Equation 1). One can determine the BED by knowing the total dose (number of fractions, n, times the dose per fraction, d) and the α/β ratio.

Equation 1:

$$BED = E/\alpha = nd\ (1+ d/(\alpha/\beta))$$

The α/β ratio is used to describe the tumor cells or late reacting cells. Rapid growing cells have a higher proliferation and are more susceptible at lower doses and have an $\alpha/\beta = 10$. Slower growing tissues, i.e. most normal cells, will be more sensitive to late effects or higher doses of radiation administered per fraction; their $\alpha/\beta = 3$.

To achieve a similar effect on tumor cells for a higher fractionation, one can solve for the total dose needed, D2, when comparing to a conventional fractionation (1.8 to 2.0 Gy= d1) to its corresponding total dose, D1.

Equation 2:

$$\frac{D2}{D1} = \frac{(\alpha/\beta + d2)}{(\alpha/\beta + d1)}$$

By using the isoeffect curves, radiation oncologists can translate this knowledge to create a clinical treatment plan for patients. Although there are certainly many factors that complicate this translation (i.e., chemotherapy and other dose-modifying agents), it serves as a starting point.

Radiation Toxicity

Although there are remarkable techniques to curb exposure of healthy tissue to radiation therapy, this has not always been the case. Early on we were able to learn more about the effects of radiation from individual accounts. One notable example was the development of aplastic anemia that affected Marie Curie after her work with radium. Even her scientific papers and home cookbook are still considered too dangerous to handle given their level of radiation exposure. On a larger scale, studies have followed the effects of radiation toxicity on victims of Hiroshima, Nagasaki, and Chernobyl over several years after the events.

Localized Clinical Syndromes

As previously mentioned, frequently replicating cells are easy targets for toxicity. Particularly, the crypt cells of the intestinal tract, bone marrow, thyroid tissue, basal cells of the dermis, oral mucosa, and salivary glands are affected. With the gastrointestinal tract, symptoms of nausea, vomiting, and diarrhea may become apparent after several treatments, especially when combined with chemotherapy. Malabsorption and electrolyte abnormalities often follow.

Radiation dermatitis may appear in a very distinct pattern consistent with the shape of the area radiated. Some accounts indicate that more than 90% of patients treated with radiation therapy for breast cancer will develop dermatitis. The lesions may be edematous, pruritic, and tender, warranting topical treatment with moisturizers and steroid creams. Long term, these lesions may remain red with evidence of thickened fibrosis. Hair growth within the affected area is commonly lost as well.

Xerostomia is found in many head and neck cancer patients after damage to the salivary glands. This causes not only dry mouth, but a significant change in the pH and bacteria colonization of the region. Poor dentition may result, and patients may develop gingivitis, periodontal disease, and dental caries despite their best efforts. Artificial saliva substitute is one agent used to help with this, as sipping on water will only help with the symptoms of dry mouth. Cholinergic agents such as pilocarpine may also stimulate saliva. Amifostine is an FDA approved agent that serves as a radiation protector for patients undergoing fractionated treatments with the intention of preserving salivary gland function.

Acute mucositis may occur when chemotherapy and radiation are given. This desquamation of the oral mucosa leaves the patient with a red, raw, and tender mouth. Depending on the severity, it may require an interruption of radiotherapy for a week while the cells regenerate. Topical anesthetics like viscous lidocaine and systemic narcotics help with the symptoms. Always examine the patient to differentiate Candida infections from mucositis and treat appropriately.

With a single dose of 2 Gy, such as in exposure from total body irradiation (TBI), or a collective fractionated dose of more than 4 Gy to the eye, as in treatment of a primary eye neoplasm, cataract development may occur. It takes several years to develop, but can easily be corrected with surgery.

TGF-β and basic fibroblast growth factor released during radiotherapy may cause acute pneumonitis and the potential for developing lung fibrosis after a prolonged latent period of months to years.

Large Field Irradiation

When administering radiation to a wide area, a reduction in the total dose is warranted. This is a commonly used technique often combined with chemotherapy for treatment of systemic disease. Total body irradiation is also used to assist in depleting hematologic tissue in preparation for bone marrow transplant. This acts to decrease the Graft-versus-Host response subsequent to allogenic bone marrow transplantation.

Whole Body Irradiation

Radiation disaster victims and those surviving exposure are an unfortunate group that have given us a greater depth of knowledge into the systemic effects of whole body irradiation. Death typically occurs with a single exposure at levels greater than 3.5 Gy. This occurs within a range of 30 to 60 days after exposure, and sooner in those patients who are unable to obtain medical assistance. The first symptoms to develop may be gastrointestinal illness with nausea, vomiting and diarrhea. Generalized muscle weakness and cognitive difficulties may also become apparent.

The cerebrovascular syndrome exhibits many of the symptoms listed above in addition to seizures and ataxia when a patient is exposed to more than 50 Gy to the entire body. Onset occurs within minutes, and death follows in 2 to 3 days. Similarly gastric illness can occur with one fraction of 8 Gy, resulting in death in less than 10 days. The bone marrow within the radiation field can be depleted with just one treatment of 2.5 Gy. Death from an increased susceptibility to infection or hemorrhage can occur as leukocytes and platelets decrease.

Radiation Induced Malignancies

In radiation disasters, as with patients undergoing radiotherapy, there is a chance of latent malignancy development. Hematologic disease may manifest itself in approximately 10 years, and solid tumors may take more than 20 years to develop.

Leukemia accounts for 20% of the secondary malignancies seen with radiotherapy. A multitude of solid tumors may develop, most of which are located within the previously irradiated field. Thyroid, breast, and salivary gland tissue are relatively radiosensitive, and are associated with a higher rate of solid tumor malignancies. Higher doses and a larger field of radiation also serve as contributing factors to the development of secondary malignancies.

Prenatal toxicity

Prenatal exposure of 0.06 Gy or more to a developing fetus may result in a wide range of abnormalities including microcephaly, neurologic defects, growth and mental retardation. Exposure within the first 8 weeks of conception most commonly leads to termination of the pregnancy. One study indicated that prenatal exposure may lead to an increased incidence of leukemia to those children born, but this was negated by those examining survivors of the Japanese atomic bomb.

Occupational Exposure

A sievert (Sv) is the measured unit for the dosage of ionizing radiation. It is a product of the absorbed dose of radiation (in Gy) and RBE. Because of the potential exposure by personnel in close contact with radiation, badges are worn to quantify the amount in Sv. The annual dose limit for the body is 5 Sv by a radiation worker. If exposure is quantified as less than 10% of this amount, then a radiation worker is not required to wear a monitoring badge.

Oxygen

As previously discussed, different tumor types and treatment regimens may have varying dose-response curves. A number of agents can be used to modify the dose-response curve in patients, including oxygen, hemoglobin, and many pharmacologic agents. As well, the blood supply providing oxygen plays a major role in tumor physiology.

The first agent discovered to modify radiation response was oxygen, as demonstrated by Gottwald Schwarz in 1909. He was able to demonstrate a decrease in skin reaction to patients undergoing compression of the blood supply while receiving radiotherapy.

It has been shown that tumors may exhibit chronic hypoxia within the deepest portion or most poorly vascularized region of a lesion. With reduced oxygen concentrations below normal tissue levels, radioresistance develops within the region. Hypoxic cells are more resistant to radiation because the lack of oxygen does not provide for a significant production of reactive oxygen species (ROS). It has been shown that lower levels of tumor oxygenation prior to radiation treatment correlate with less cell kill.

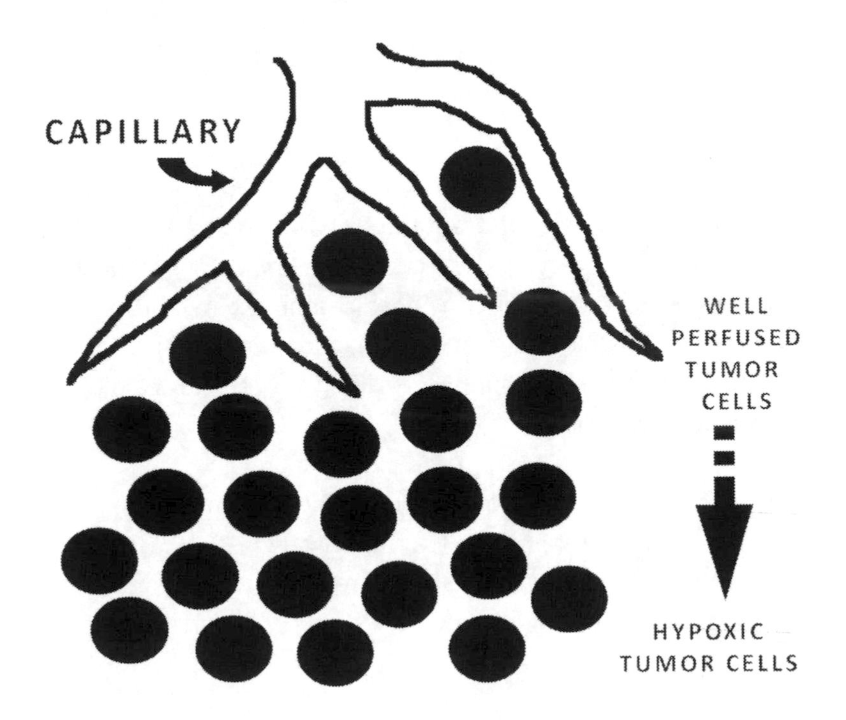

Tumor blood flow may also be erratic, subjecting certain portions of a tumor to intermittent periods of hypoxia, a concept known as acute hypoxia. As blood flow may fluctuate within a tumor, many agents may be used to alter oxygen, hemoglobin, or the blood vessels themselves to minimize the hypoxia during radiotherapy treatments.

Hypoxia within a tumor may vary in size, affecting as little as 1% of the lesion, or as much as half. Measuring hypoxia is difficult, and no one mechanism is precise. The most popular instrument used to indirectly measure hypoxia is electrodes placed directly into the tumor. This was first tested in head and neck cancer patients. The measurements of tumor oxygen concentrations were weighed against overall survival, and patients with lower pO_2 levels were found to have worse results.

A number of studies have shown that increasing oxygen levels to tumor sites, thus decreasing hypoxia, have a beneficial effect. The methods to carry out such a task have been numerous.

Decreasing Hypoxia

Increase Inhaled O_2

One of the most obvious ways of reducing hypoxia is simply breathing higher than normal levels of oxygen during radiotherapy. A number of studies were performed with a gas known as carbogen (a mixture of 95% oxygen and 5% carbon dioxide) delivered under hyperbaric conditions. The hyperbaric chambers were later found to yield no difference than gas inhaled under normal atmospheric conditions and were phased out. Patients also complained that the hyperbaric conditions were uncomfortable. After numerous trials, there did not appear to be any significant improvement in tumor control.

Hemoglobin

As shown with hypoxia, lower patient hemoglobin levels affect the tumor response to radiotherapy. A patient with lower hemoglobin levels is expected to have a decreased response to radiotherapy. Methods to increase hemoglobin and to determine what level have been the subject of much research. Having a hemoglobin level equal to or greater than 11g/dL were shown to improve survival. Some patients needed to undergo blood transfusions to meet this target level, and the result yielded a higher oxygen level within the tumor as demonstrated with probing electrodes.

Another method thought to raise hemoglobin is by administering erythropoietin. Very few studies have been done, and there is great controversy that erythropoietin, being a hormone that stimulates cell division, may actually grow tumor in the process. This is largely theoretical and has not been demonstrated.

Pharmacologic agents

As previously discussed, acute and chronic hypoxia play a major role in the delivery of oxygen to the tumor site. Nicotinamide, a derivative of vitamin B_3, is a medication thought to assist with varying the blood supply to an area of tumor that may be acutely or chronically hypoxic. Few trials with this medication are complete, but the initial results are encouraging.

A class of pharmacologic agents known as nitroaromatic compounds has been found to seek out areas of hypoxia and have a high electron affinity. Several trials have been performed examining their effectiveness with chemotherapy and radiation. One of these nitroaromatic compounds is metronidazole, also known for its antimicrobial properties against anaerobic organisms (thus demonstrating its effectiveness under hypoxic conditions). Another member of this nitroaromatic family is misonidazole which, when combined with radiotherapy, did not show much benefit among several studies including the Danish Head and Neck Cancer study 2 (DAHANCA2). However, nimorizole was used for the DAHANCA5 study, and exhibited significantly better local tumor control when combined with radiotherapy. Nimorizole and other nitroaromatic compounds still require further trials, but nimorizole is already standard therapy in Denmark to treat head and neck tumors.

In theory, use of the nitroimidazole agents may help to overcome the effects of hypoxia as each round of radiotherapy is administered. However, studies have demonstrated that despite cell kill in hypoxic regions of a tumor after treatment with nitroimidazoles, as well as other agents, tumor cells will produce new areas of hypoxia. This phenomenon is known as rehypoxiation. As each new cycle of radiotherapy is given, this newly produced hypoxic region of the tumor is by definition more radioresistant than before.

> Neurotoxicity, notably peripheral neuropathy, is the greatest side effect with the use of nitroimidazole therapy.

Pentoxifylline (Trental™) has been used in patients with severe vascular disease because it produces a greater amount of flexibility within the red blood cell membrane. This allows the cell into smaller capillaries, promoting the delivery of oxygen to previously hypoxic regions. Its effects work well when given at the time of radiation treatment, but fail to produce any benefit thereafter when administered on a daily basis.

Perflourocarbons are molecules known as "artificial blood substitutes" that are able to carry oxygen in the blood. No evidence on their effectiveness has been published, but they remain an ongoing area of interest.

Tumor Vascular Supply

Without an adequate amount of blood flow, no cell can persist for very long. The nutrients provided by the vascular supply may be small in quantity under hypoxic conditions, but without them cell death and necrosis prevail. Thus, the importance of targeting the vascular supply has been a major focus in cancer research.

Because tumor cells themselves are erratic, their microvascular supply may also exhibit these features. Blood vessels in tumor pathology are more disorganized than normal vessels and prone to leakage. This is thought to be the result of the abnormal cytokine production by the tumor cell that is activating local angiogenesis.

Hypoxia induced factor-1A (HIF1A) is a transcription factor that regulates the pathway that induces angiogenesis. HIF1A is produced when hypoxia is present within the poorly vascularized, and thus poorly oxygenated, core region of a tumor. These severely hypoxic tumor cells are then better equipped to carry out their life cycle and replication by adapting to the circumstances. Unfortunately, this will allow for these radioresistant hypoxic cells deep within the tumor to continue to multiply.

There are many other potential molecules involved in the process of angiogenesis that have been the focus of new therapeutics including angiostatin, interferon α and β, endostatin, and thrombospondin. The use of these agents in combination with radiotherapy is still ongoing.

Pharmacokinetics

The goal of both chemotherapy and radiotherapy is to create the greatest amount of toxicity for tumor cells while minimizing the amount of damage done to normal tissue in the process.

The Goldie-Coldman hypothesis states that some tumor cells within a lesion are more sensitive to a certain agent of therapy, whereas others may be rather radioresistant to that same agent. This also follows the idea of tumor heterogeneity, that within a tumor genetic instability arises and more mutations are created as the tumor proliferates. Several tumor cells with various genetic makeup are now present, all of which will have a different response to a chemotherapeutic or radiotherapeutic treatment regimen.

There are a few definitions to describe the interaction between two or more agents given to combat tumor cells:

- *Zero interaction* is the term used to describe two treatments that function independently of one another and produce the expected additive results. If these agents had been given separately, they would yield no more or less cell kill.

Example – Radiation kills 2 logs of tumor cells and chemotherapy kills 3 logs of tumor cells. As a result the tumor exhibits a total of 5 logs of cell kill, yielding zero interaction between the 2 agents.

- *Antagonism* is the worst result possible when combining two agents, as the interaction between them would actually yield less cell kill combined than if each were used independently.

- *Positive interaction* is finding that the two agents together allow for greater cell kill when given together than if each were given separately. It is important not to misuse the word synergy or synergism when referring to this interaction. Note that when combining chemotherapeutic agents, it is necessary to decrease the dose from that of the individual agent that may be given alone so as to compensate for side effects and normal tissue toxicity.

- *Synergism* implies the same concept of positive interaction, but kinetic data must confirm this with use of complex mathematical models such as an isobologram analysis.

> An isobologram is a model that uses the cytotoxicity of two different agents and allows for a computation that will determine if their interaction is additive, antagonistic, or synergistic.

A further insight into the interaction of various agents can be examined by applying the *median effect principle analysis* as created by Chou and Talalay and is based on the properties of enzyme kinetics. This mathematical model may determine if two agents are mutually exclusive or mutually nonexclusive. Mutually exclusive agents carry out their cytotoxicity in a similar manner, whereas mutually nonexclusive agents act independently of one another. An additional concept known as the combination index (CI) allows quantitative determination of drug interactions, where CI < 1 (a synergistic effect), equal to 1 (an additive effect), and > 1 (an antagonistic effect).

Despite the best analysis, there is terrible difficulty in taking data from the laboratory and applying it to the clinical setting. How an agent or group of agents react with cells *in vitro* and *in vivo* does not always indicate that it will lead to a similar result when carried out in clinical trials. One important example is the treatment developed for anal cancer in the 1970's before much of this analysis was available. The use of mitomycin C and 5-FU in combination with radiotherapy was introduced over thirty years ago, yet it remains standard treatment. Despite isobologram analysis indicating mixed results, clinical trials have proven this regimen to work best.

Spatial cooperation is the concept that describes the interaction between chemotherapy and radiation. Chemotherapy can be used to kill the cells outside of the region of a tumor that is not involved in radiation therapy. Most often this would imply metastatic disease. However, there are some regions that chemotherapy may not be able to penetrate, thus radiation is left to act as the sole cytotoxic agent. With this kind of interaction, there is no assumption that the chemotherapy and radiation are working in an additive manner, there is merely a focus on greater disease control.

Chemotherapy is a radiosensitizer, making tumor tissue more susceptible to cell kill when radiation is used. This allows for a decrease in the dose of radiation to promote the same amount of cell death. The taxane based chemotherapeutics target the spindle apparatus, whereas the others may affect the DNA repair mechanism.

Chemotherapy agents

As discussed, a chemotherapy agent may act as a radiosensitizer. However, an agent may work with radiation by other means.

For example, Tirapazamine (SR-4233) has not shown any significant radiosensitizing effects. Tirapazamine is an agent that specifically targets anoxic cells over well-oxygenated cells and kills them. Tirapazamine produced more DNA strand breaks under anoxic conditions when compared with aerobic conditions in squamous cell carcinoma cell lines. When combined with cisplatin, an agent known for its toxicity on well-oxygenated cells, tirapazamine has been shown to exhibit good results. This allows for both agents working together to target all tumor cells, hypoxic or otherwise.

The Biologics

Traditionally, it was the chemotherapy agents that were used to fight cancer, most of them with cytotoxic properties causing harsh side effects. While many of them remain the gold standard of treatment, newer agents with an entirely different approach are being introduced. These novel agents may work primarily with radiation, as seen by the radioimmunologics. Others have varying radiosensitizing properties, and some are being used without radiation at all. One of the goals of these emerging new regimens is to create a safer side effect profile with less toxicity.

> The definition of a *biological* agent, with respect to pharmacology, is a therapeutic agent that directly targets a certain molecule. A number of molecules regulating cell growth and apoptosis have been the aim of biologics.

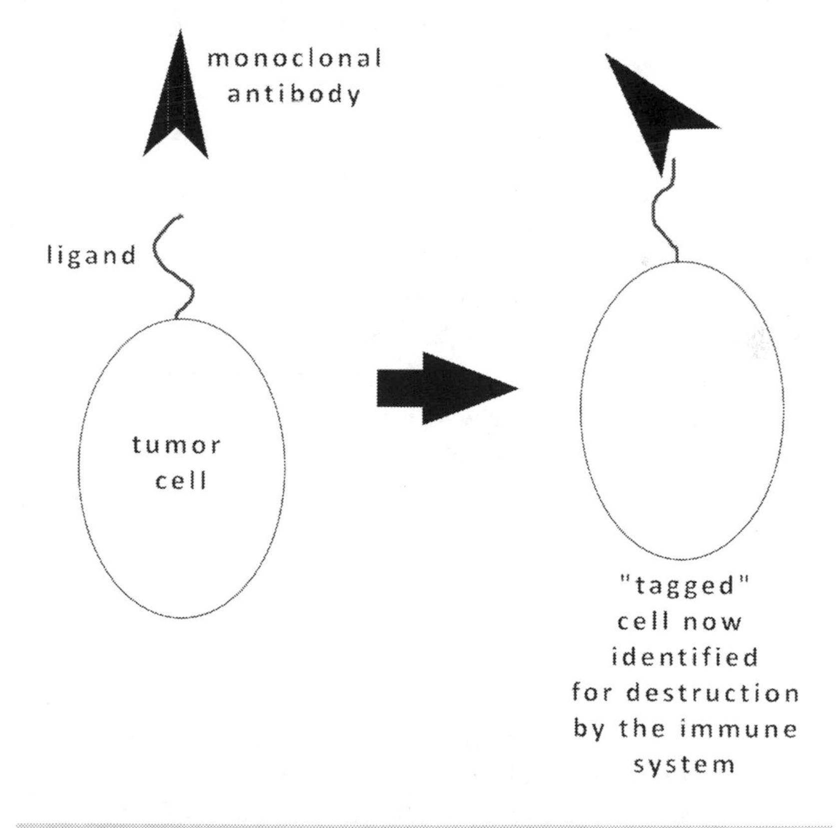

A biologic agent such as a monoclonal antibody targets a tumor cell that exhibits a specific cell membrane ligand. The targeted cell is then marked for destruction

Epidermal growth factor receptors are a group of four separate proteins-EGFR, HER2, HER3, and HER4-that function as tyrosine kinase receptors. These proteins span the cell membrane with surface ligands to the extracellular space, as well as a portion within the intracellular matrix. When targeted by a biologic agent, epidermal growth factor receptors undergo a conformational change that activates intracellular tyrosine kinase. This enzyme then phosphorylates selected substrates within the cell that transduce the signal to the nucleus and, in the case of tumor cells, alter oncogenic expression.

EGFR

Cetuximab, as known as Erbitux ™, is a monoclonal antibody that targets the epidermal growth factor receptor (EGFR) ligand found on cell membranes. Cetuximab has been shown to work well with other chemotherapeutic agents in combination therapy against malignancy when given on a weekly basis. It has been shown to exhibit both chemotherapeutic and radiosensitizing properties. EGFR is thought to play a major role in tumor growth in various forms of cancer, notably gastrointestinal malignancies such as pancreatic carcinoma. Cetuximab is approved for irinotecan-refractory colon cancer as single agent therapy and has been shown to work in combination with radiation for treatment of head and neck SCC in patients who have had prior platinum-based therapy. Unfortunately, some acquired resistance to cetuximab treatment has been noted. With respect to its toxicity, up to 3% percent of patients receiving cetuximab were noted to have an anaphylactic reaction which can be fatal.

Erlotinib (Tarceva™) is a tyrosine kinase inhibitor of EGFR used in non-small cell lung cancer, pancreatic cancer, and several other malignancies. Side effects include skin reaction and diarrhea.

Gefitinib, as known as Iressa ™, works similarly to erlotinib and is approved for non-small cell lung cancer after prior chemotherapy treatments have been given, but a 2004 study failed to demonstrate a survival advantage in these patients. The most common side effect appears to be acne.

HER2

Trastuzumab (Herceptin ™) was the first monoclonal antibody introduced and is approved for HER2 positive breast cancer. This agent functions by binding to the HER2 protein ligand on the cell membrane. Binding of trastuzumab to the HER2 receptor creates an immune-mediated response against the now "recognized" cell, initiating cell death.

Twenty to thirty percent of all breast cancers exhibit HER2 overexpression, and testing for this receptor is typically performed at the time of surgical tissue biopsy. For the remaining majority of breast malignancies, however, trastuzumab is not indicated. One unforeseen side effect of trastuzumab has been cardiotoxicity with a significant amount of congestive heart failure. When using radiation and trastuzumab in combination, the effects of cardio- and neurotoxicity have been said to be increased, but there is no data to confirm this.

With respect to radiotherapy, it should be noted that tumor cells overexpressing HER2 have been shown to be more radioresistant. This may also be the case for cells demonstrating an upregulation of EGFR expression. By applying biologic therapy, these same tumor cell lines will exhibit a modest component of radiosensitization. As well, it has been demonstrated that EFGR inhibitors can decrease tumor angiogenesis.

Fortunately, lymphoma is a very radiosensitive malignancy. The B cell agonist rituximab (Rituxan™) is commonly used with radiotherapy for treatment. Though known as an effective agent for rheumatologic disease, rituximab is also approved for B cell non-Hodgkin lymphoma. Rituximab works by inducing apoptosis of cells expressing the CD 20 ligand. Because it acts as an immunosuppressant clearing B cells from the body, patients may become more susceptible to infection. More notes on rituximab are featured in the radioimmunology section of this book.

Vascular endothelial growth factor

Vascular endothelial growth factor (VEGF) is found to be the molecule responsible for mediating much of the blood vessel growth used to help vascularize tumor tissue. These tumor blood vessels that are created are often poorly formed and chaotic in structure when compared to those in healthy tissue. They may exhibit increased permeability and create areas of hypoxia within the tumor mass. When hypoxic, NF-κB, an important pro-angiogenic molecule to be discussed later, is also upregulated. VEGF expression has been seen in a variety of malignancies and correlates with a worse clinical prognosis.

Angiogenesis Inhibitors

To cripple blood vessel proliferation within the tumor, a number of different approaches have been used:

1. Target the poorly formed blood vessels. Their inherently weakened composition makes them very susceptible to destruction.
2. Inhibit VEGF with therapeutics known as vascular targeting agents (VTA).
3. Inhibit tyrosine kinases responsible for the cascade of cell signaling that results in VEGF receptor activation.

Bevacizumab (Avastin™) was the first monoclonal antibody targeting VEGF available in the United States. It currently has FDA approvals for colon cancer, non-small cell lung cancer, and breast cancer (despite reports that it may not prolong survival in breast cancer patients). Side effects include gastrointestinal (GI) bleeding and perforation in patients being treated for GI malignancies. Because bevacizumab also limits the production of new blood vessels, it may also cause higher blood pressure in some recipients.

Vandetanib or ZD6474 (Zactima™) is a newer agent with anti-EGFR as well as anti-VEGF receptor properties being studied in clinical trials with non-small cell lung cancer.

ZD6126 is a prodrug of an older medication, colchicine. It works to inhibit microtubule polymerization by binding to tubulin, the main component of microtubules. Availability of tubulin is necessary for mitosis, therefore inhibiting cell division.

NF-κB

NF-κB is a known transcription regulator present in many healthy and tumor cells that works to inhibit apoptosis. Research has shown this molecule to be responsible for treatment resistance in many cancer patients receiving both chemo- and radiotherapy. Thus, the pathway activating NF-κB has been a focus of potential treatment options.

Bound to I-κB in the cytoplasm of the cell, NF-κB is released when activated and migrates to the nucleus where it binds to various anti-apoptotic genes including BCL2. Increasing the amount of I-κB within cells to bind with NF-κB is thought to be one mechanism of treatment.

Since I-κB is destroyed by enzymatic proteasomes, some medications like bortezomib (Velcade™) work to inhibit this proteasome activity. With a greater amount of I-κB within the cell, more NF-κB is bound and inactive. Bortezomib is being used in multiple myeloma patients that are refractory to chemotherapy treatment. Side effects include marrow suppression or gastrointestinal symptoms, and possibly neuropathy. Although well tolerated, this agent has a very narrow therapeutic index.

Radiation has been said to induce NF-κB activity, and its expression correlates with radioresistance. By inhibiting NF-κB and administering radiation, apoptosis flourishes within tumors previously thought to be radioresistant.

Tumor necrosis factor related apoptosis inducing ligand (TRAIL) is a transmembrane protein found in many normal tissues as well as tumor cells. Because activation of the TRAIL pathway results in apoptosis in tumor cells with a minimal effect on healthy cells, it has the potential to be a very selective agent. TRAIL is activated by the binding of proteins DR4 and DR5. Laboratory work on these proteins indicate that inactivation by monoclonal antibodies increases the cytotoxicity due to TRAIL. As a result of these findings, much research has focused on producing monoclonal antibodies to each. TRA-8 is found to target DR4 and 2E12 is found to target DR5. Studies focusing on combining TRA-8 and 2E12 monoclonal antibodies with chemotherapy and radiation are now being carried out on various cancer cell lines.

The future of biologics

The initial use and FDA approval of many of these newer agents has been in patients with refractory disease to more traditional chemotherapeutic regimens. As their effectiveness is evaluated in clinical trials, they may prove (or in some case have already become) to be a part of first-line therapy in the future.

Work combining monoclonal antibodies and more traditional chemotherapy agents, in addition to radiotherapy, remains the focus of much laboratory and clinical trials.

Radioimmunotherapy

90-Yttrium ibritumomab and 131-Iodine tositumomab are two well studied agents used in non-Hodgkin lymphoma (NHL). Both are monoclonal antibodies to the CD 20 ligand, which is only found on B cells. Recall, rituximab (Rituxan™) is also an IgG1 monoclonal antibody specifically targeting CD 20. What sets these other 2 agents apart is their attachment to radioactive elements.

Radiotherapy has been an essential tool in combating NHL, as lymphomas have proven to be very radiosensitive. It has been used alone and in combination with chemotherapeutic agents. Advantages of radiotherapy include the targeting of occult tumor cells and the lack of increased risk of secondary hematologic malignancy.

The combination of the monoclonal antibody and the associated radioactive element allows for a dynamic process known as the crossfire effect. A radiolabeled monoclonal antibody is able to attach itself to a B cell. The next step allows for the radioactive element to kill the attached B cell. These agents are much more useful in bulky or poorly vascularized tumors that would not otherwise allow for adequate chemotherapy penetration.

Studies measuring the effectiveness of ibritumomab and tositumomab both examined patients with refractory or relapsed B cell NHL, absolute neutrophil count >1500, platelets >100K, and with less than 25% of bone marrow involvement.

Ibritumomab is a monoclonal antibody attached to 90-Yttrium (Zevalin™) first introduced in 2002. The efficacy of ibritumomab was first demonstrated when patients were enrolled in a study administering rituximab with and without ibritumomab. A significant overall response rate in patients receiving combination therapy was shown at 80%, compared to 56% with rituximab alone. Biodistribution of the molecule was noted to be highest in the areas of the axillary, inguinal, and periaortic lymph nodes as well as the spleen and liver.

Tositumomab is a monoclonal antibody attached to 131-Iodine (Bexxar™). Tositumomab was shown to work better in patients with a smaller tumor burden (<500 grams), no prior radiotreatments, positive bone marrow involvement, and lower grade histology.

Given the data from tositumomab and ibritumomab, a registered indication for use in relapsed/refractory indolent or transformed lymphoma was approved. Comparing the agents, certain individuals may benefit from one agent over another. There does not appear to be a difference in the complete response rate between both agents at 12 weeks, nor any difference in overall survival. However, these numbers have been widely questioned as the study enrolled only 30 patients.

Whereas both agents were well tolerated, myelosuppression was the most notable side effect. Bexxar™ proved to have a less severe effect on the decline in platelets, thus making it the agent of choice in patients with lower initial platelet counts. Other differences are noted in the table:

Agent	Type of decay	Type of Monoclonal Antibody	Half -life	Particle Energy (MeV)	Particle length (mm)
ibritumomab	β radiation	IgG1	64 hrs	2.293	5.3
tositumomab	β and γ radiation	IgG2a	193 hrs	β 0.606 γ 0.364	0.8

Notice the increased particle length of Zevalin™, thus allowing it to work better on patients with bulky tumor burden, better penetrating the lesion. Zevalin™ was noted to have associated hepatotoxicity and should be avoided in patients with liver disease. Bexxar™ has an associated cardiotoxicity, and prior cardiac workup maybe warranted prior to treatment with this agent.

Further studies have noted that repeated treatments of these radioimmunotherapeutic agents are not only safe, but also effective with complete response rates greater than 50%. Of note, myelosuppression did remain an issue with these patients undergoing re-treatment.

The potential to develop AML and other therapy-related myelodysplastic syndromes is a concern anytime chemotherapeutic agents are used. Radioimmunotherapy was noted not to cause any increased risk of these complications when compared to chemotherapy alone.

Gene Therapy

The next step in dose modifying agents may be applying genetic sequences via transfection into the cell to alter the susceptibility of tumor cells to chemotherapy. This has been applied in studies for which the gene for cytosine deaminase was given to tumor cells to assist in breaking down 5-FC to its active metabolite 5-FU in an effort to induce further cytotoxicity. The technology of transfection delivery via naked DNA, lipoplexes, and other molecules serving as vectors for gene sequences is continually improving.

SECTION TWO:

PHYSICS

“Most residency programs in the country teach a formal physics course. Physicians work with physicists to ensure that treatment is delivered accurately to patients. Though you don't have to have a love affair with the subject material, you should at least be able to understand the material."

-Association of Residents in Radiation Oncology (ARRO), arro.org

Subatomic Particles

An overview of physics must begin with reviewing the definitions and basic terminology.

The number of protons in an atom is represented by the atomic number, Z, and the number of neutrons is represented by the neutron number, N. The mass number, A, is a combination of Z plus N. Note that the atomic mass is merely an estimate, and that other subparticles with minute amounts of mass also have value. Thus, the actual nuclear mass differs slightly from the atomic mass.

Recall that the atomic mass of Carbon is 12. The atomic mass unit (amu) is $1/12^{th}$ the weight of the Carbon nucleus, comprised of 6 neutrons and 6 protons. Neutrons and protons comprise the nucleus and are referred to as nucleons. The proton has a +1 charge, and the neutron has no charge.

> An isotope is an atom characterized by the number of neutrons (and thus differing atomic weights). For example, Carbon-12 and Carbon-14.

Other, less familiar particles have been theorized over the last few decades. A quark has been found to be the elementary unit of a nucleon. A quark may carry either a positive or negative charge of +1/3 or -1/3. Quarks are classified by six different flavors, the most common of which are up and down. The other flavors are charm, strange, top, and bottom, all of which decay rapidly and are only seen under rare circumstances. All elementary particles have angular momentum as characterized by their spin. Quarks and electrons have a spin of ½.

Protons are comprised of 2 up quarks and 1 down quark

Neutrons are comprised of 1 up quark and 2 down quarks

The neutrino is a neutral particle emitted as part of beta decay and has a spin of ½.

The electron has a -1 charge, but the atomic mass is 1/2000th of a nucleon.

Atomic Structure and Charges

The older concept of protons and neutrons comprising the nucleus of an atom with orbiting electrons has been theorized to be a much more fluid model than previous described. There are at least 160 different types of particles that have been described to comprise what we know as matter. Electrons fill shells around the nucleus, each of which is comprised of defined energy levels.

Electron shells are denoted by *n*, the *principle quantum number.* For the figure on the following page, the *k* shell has a value of n=1. The *l* shell has a n=2, and the *m* shell has a n=3. Only a certain number of electrons may occupy each orbital as dictated by the formula $2n^2$. Thus, 2 electrons may occupy the *k* shell and 8 electrons may occupy the *l* shell.

Electrons fill the *k* shell first, followed by the *l*, and then the *m* and continue outward to a potentially infinite number of shells. The *k* shell is considered the lowest energy level, and energy levels increase among each orbital as they expand outward. This energy level is not a subjective value, but instead there is a defined energy state for every orbital.

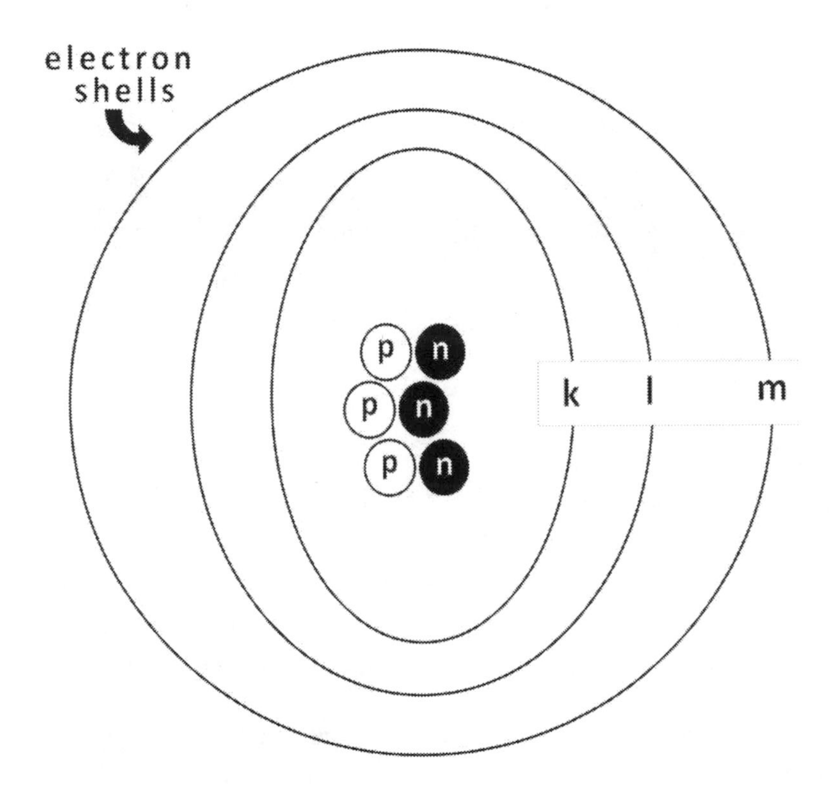

An atom may react with two possible outcomes when energy from outside the system is applied. *Excitation* of an atom occurs when an electron is displaced from an inner orbital to that of an outer orbital, whereas *ionization* occurs when the electron is completely removed from the atom. The amount of energy holding the electron to the atom has been overcome in ionization, and the charge of the atom is now changed to a more positive state.

In a similar manner to excitation, consider the situation of an electron from a higher energy shell *m* moving into a lower energy shell *k*. Now we will assign a value of 5 to represent the energy necessary for an electron to exist in shell *m*. Similarly, we will assign an energy level of 1 to the lower *k* shell. When the electron moves from the *m* to the *k* shell, the 4 units of energy between the levels is released in the form of photons. The number of photons released is referred to as the *fluorescent yield*. As an electron transitions from a much higher *n* level to a lower *n* level, a greater amount of energy is released.

When an electron is removed from a lower orbital, an electron from a higher shell may fill the vacancy, resulting in a release of energy. This energy may be released in the form of a photon as described above, however it may also be transferred to another electron, termed the *Auger electron* (discussed later), which is ejected from the atom.

The *positron* is the antimatter counterpart to the electron. It is similar in weight, but opposite in charge. When a positron and an electron combine via a process called the *annihilation reaction*, two photons are produced travelling in opposite directions, yielding 0.51 MeV (millielectron volts) each.

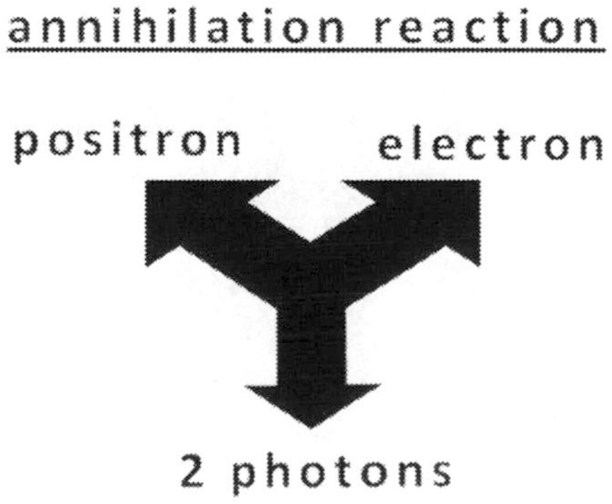

annihilation reaction

positron electron

2 photons

The reverse reaction may also occur and is referred to as *pair production*. In this process a photon with an energy value of 1.02 MeV or greater is transformed into an electron and a position. Each carries with it a value of 0.51 MeV in addition to half the value of the energy greater than 1.02 MeV from the originating photon.

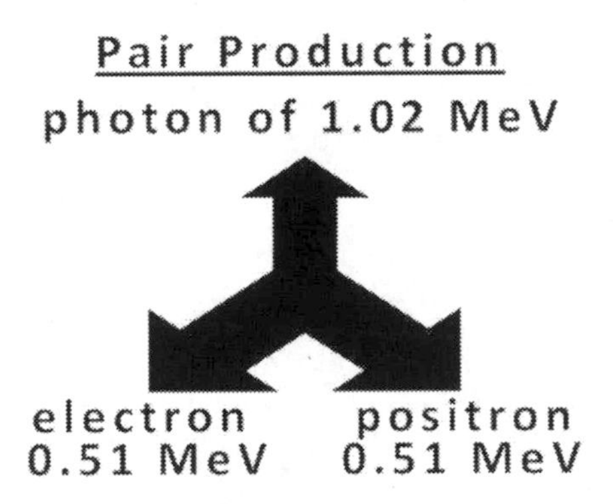

Forces of the universe

Matter has been shown to exhibit four different forces that act upon itself within the universe. Each transmits a respective particle and carries a specific amount of energy.

Strong Charge transmits the gluon particle and is considered the strongest of the four charges. It is the standard upon which other charges are compared, carrying a relative strength of 1. It is the force that binds protons and neutrons together within the nucleus.

Weak charges are transmitted by the W and the Z particle with a relative strength of 10^{-6}.

Electric or *electromagnetic charge* is transmitted by electrons and quantified by the Coulomb with a relative strength of 1/137th that of the strong charge.

Gravity has the smallest strength at 10^{-37} and is transmitted by the graviton.

Radiation Types

Two main types of radiation exist: Ionizing and Non-ionizing. To ionize an atom, the energy delivered must be strong enough to displace electrons from their shells. This may be accomplished directly or indirectly. Direct ionization occurs with charged particles, whereas indirect ionization uses neutral particles.

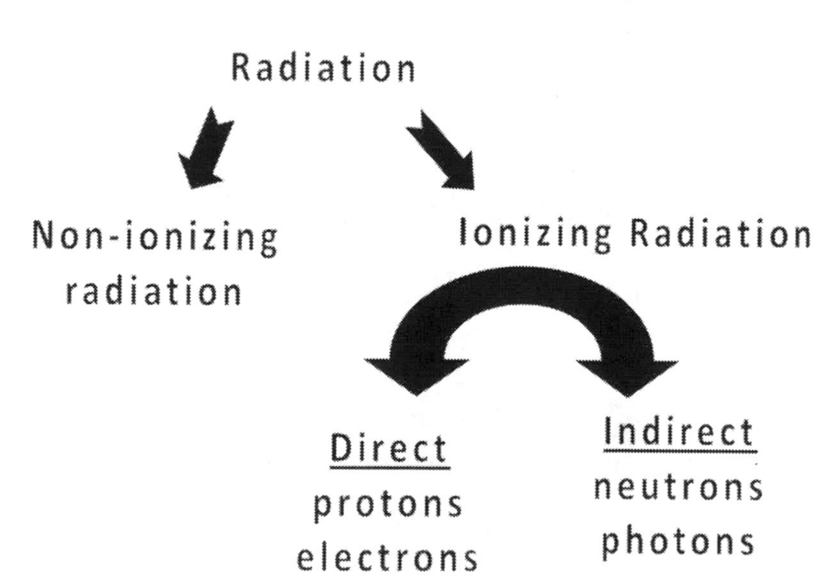

For radiation therapy to be effective, we must concentrate on the use of ionizing forms of radiation such as x-rays and gamma rays. If the radiation is non-ionizing, then no ions within the biological tissue would be created, and no significant cell damage would occur.

Diagnostic vs. Therapeutic radiation

It may be difficult to understand how x-rays are created for simple procedures like a diagnostic film and again used for treatment of malignant lesions. A diagnostic x-ray is created by a vacuum tube that uses energy to accelerate electrons to a high velocity. The electrons collide with a metal target, usually tungsten, which creates the desired x-rays. The energy from the machine is measured in volts (V), whereas the x-rays produced are measured in electron volts (eV).

A typical AAA battery is 1.5 volts.

Diagnostic plain film x-rays are usually produced by a machine with less than 150kV of energy. This means that 150 kV are supplied to the machine and electrons are accelerated to produce x-rays of 150 keV or less. To put this in the context of ionization, Cesium only requires 4 eV of energy to remove an electron from its outer shell.

By comparison, the linear accelerator (covered in the next section) used in radiation oncology for therapeutic treatment of tumor tissue creates a beam of high energy electrons, typically at a level of 4 MeV.

Nuclear Reactions

Nuclear reactions involve two entities, an atom and a projectile. The nucleus of the atom is hit with the projectile thus altering the composition of the nucleus. The change in the nucleus results in an unstable product. This product then undergoes radioactive decay, and will continue to do so until a stable nuclear species can be produced.

The half-life is the time ($t_{1/2}$) attributed for half of the quantity of a radioactive material to decay from its initial amount. Each radioactive substance has a decay constant (noted by the symbol λ) and is derived by the half-life in the equation:

$$t_{1/2} = \frac{\ln 2}{\lambda}$$

Where ln 2 is the mathematical constant equal to 0.693147181.

Types of radioactive decay

α Alpha decay

An alpha particle consists of 2 neutrons and 2 protons: ${}^{4}_{2}He$, which is a Helium nucleus. As you might expect, this is a relatively heavy and large decay product. An example is the decay of uranium:

$${}^{238}U \rightarrow {}^{234}Th + {}^{4}_{2}He$$

The alpha particle has a kinetic energy of 5 MeV and a speed of 15,000 km/sec. They are likely to lose their energy within a few centimeters after being emitted. A layer of skin easily blocks an alpha particle. Only through ingestion or inhalation of a formidable dose of alpha particles can radiation sickness occur.

> Note: The scandalous 2006 death of former Russian K.G.B. agent Alexander Litvinenko was attributed to radiation poisoning by the alpha emitter Polonium-210. A large quantity was found in his body, and he was buried in a lead lined coffin.

β Beta decay

Similar to the electron and its antimatter component the positron, there is another elementary particle called the *neutrino* and *antineutrino*. Both of these can be seen as products of beta decay.

In β^- decay a neutron is transformed into a proton, releasing an antineutrino and an electron as demonstrated by the reaction when Cobalt-60 is transformed into Nickel-60:

$$^{60}_{27}\text{Cobalt} \rightarrow {}^{60}_{28}\text{Nickel} + e^- + \text{anti-neutrino}$$

In β^+ decay a neutron is transformed into a proton, releasing a neutrino and a positron as demonstrated by the reaction when Nitrogen-13 is transformed into Carbon-13:

$$^{13}_{7}\text{Nitrogen} \rightarrow {}^{13}_{6}\text{Carbon} + e^{+} + \text{neutrino}$$

Some nuclei can undergo *double beta decay* (also noted as: ββ decay) whereby two neutrons of the nucleus are converted to protons. This also allows for two electrons and two anti-neutrinos to be emitted as seen with germanium-76:

$$^{76}_{32}\text{Ge} \rightarrow {}^{76}_{34}\text{Selenium} + 2e^{-} + 2\ \text{anti-neutrino}$$

The neutrinos produced have near zero mass and travel at speeds close to the speed of light. They come in *flavors*, similar to quarks, and are plentiful in our environment. A byproduct of nuclear fusion from the sun, more than 50 trillion neutrinos pass through the human body every second.

Electron capture

In this process, also known as *inverse beta decay*, an electron from the atom's innermost *k* shell is absorbed into the nucleus with a proton. A neutron is produced and a neutrino is released. As a result, the Auger effect occurs, and as an electron from a higher shell fills the empty slot within the *k* shell.

proton + electron = neutron + neutrino + Auger effect (with photon release)

An example:

$$^{26}_{13}\text{Aluminum} + \text{electron} \rightarrow$$

$$^{26}_{12}\text{Magnesium} + \text{neutrino} + \text{Auger photon}$$

Double electron capture is an advanced and rare form of decay that may occur with krypton as demonstrated:

$$^{78}_{36}\text{Krypton} + 2\ \text{electrons} \rightarrow$$

$$^{78}_{34}\text{Selenium} + 2\ \text{neutrinos}$$

γ Gamma decay

Gamma rays are the highest frequency electromagnetic wave, exhibiting the shortest wavelength. It is the most damaging form of ionizing radiation to biological tissue, easily penetrating the skin of humans. Most of the damage incurred involves double strand DNA breaks. This allows for its use in the sterilization of medical equipment. Moreover, gamma rays have also been adopted by radiation oncologists wielding the *gamma knife*, which we will cover in the next section.

One of the striking differences between x-rays and gamma rays are that gamma rays are created from changes that occur within the nucleus of an isotope, whereas x-rays originate from energy changes of peripheral electrons. Usually, beta decay precedes the release of a gamma ray. For example, cesium-137 first undergoes the following reaction:

$$^{137}_{55}\text{Cesium} \rightarrow \, ^{137}_{56}\text{Barium}^* + e^- + \text{anti-neutrino}$$

The newly formed barium-137 nucleus is unstable, and has a half life of only 2.55 minutes. It will undergo another reaction releasing a gamma ray from the nucleus without altering the number of protons or neutrons:

$$^{137}_{56}\text{Barium}^* \rightarrow \, ^{137}_{56}\text{Barium} + \gamma \text{ ray}$$

Cluster decay

When a cluster of neutrons or protons are emitted from the nucleus, *cluster decay* is said to occur. This is a rare event, as in the case of radium-223 which typically undergoes alpha decay. However, after roughly one billion alpha decays, a cluster decay occurs. The radium isotope releases a carbon nucleus in the following reaction:

Radium-223 → Carbon-14 + Lead-209

Photon interactions

When a photon beam is directed at a target, one of two actions may occur. Some photons enter the material and exit without any change in their energy. These are referred to as *transmitting* photons. However, when the photons interact they are said to be *attenuating*. It is the attenuating photons that cause ionization of atoms.

We have already reviewed pair production and the annihilation reaction. There are five interactions that occur between atoms and photons:

1. Coherent scattering
2. The photoelectric effect
3. The Compton effect
4. Pair production & annihilation
5. Photodisintegration

Coherent scattering accounts for a very small amount of photon interactions. In this process, an incoming photon hits an orbital electron, most likely one in an outer shell. The electron and the incoming photon may change directions, but no energy is traded from one to the other. This change in direction from both components separates the process from that of a *transmitting* beam. Because this process occurs with photons of such low energy levels, typically less than 10 keV, no ionization occurs as the electron does not leave its orbital.

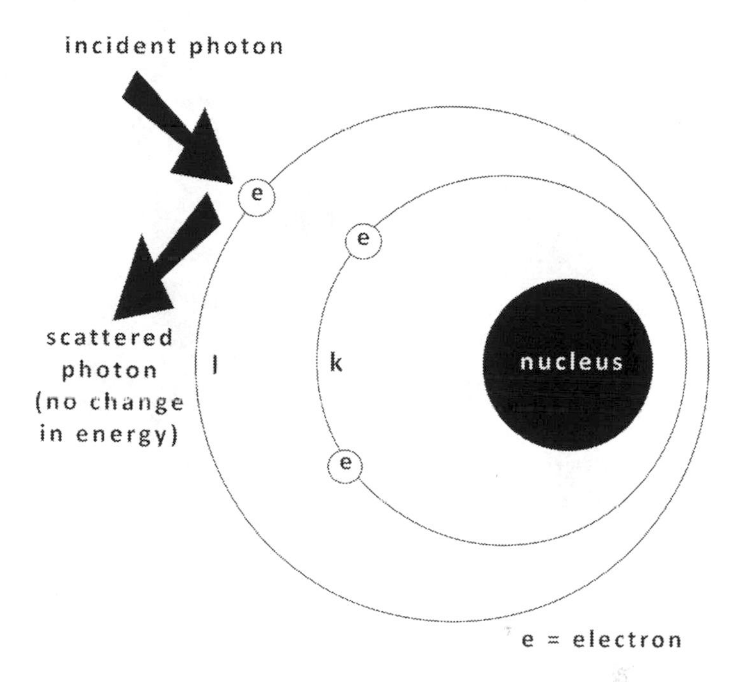

In the photoelectric effect, an incoming photon interacts with an electron, usually from the innermost *k* shell. The photon and its associated energy are taken in by the electron, producing a *photoelectron*. This high energy electron is then ejected from the atom. An electron from a higher shell then drops into the vacant lower energy shell. This may release an Auger electron or a photon. This process may occur repeatedly as the electrons continue to drop down to fill in the vacant spots of the lower electron shells.

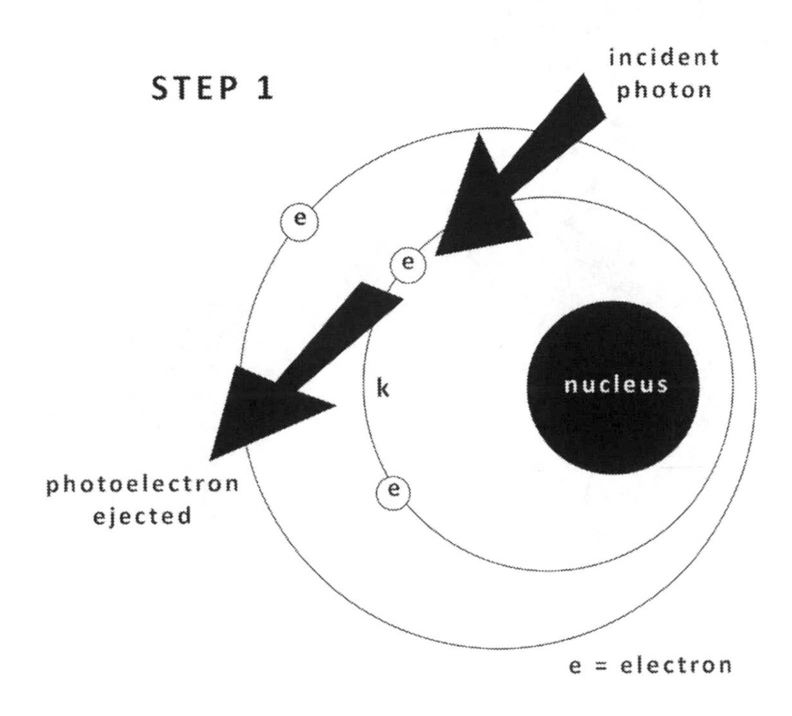
STEP 1
incident
photon
e
e
k
nucleus
photoelectron
ejected
e
e = electron

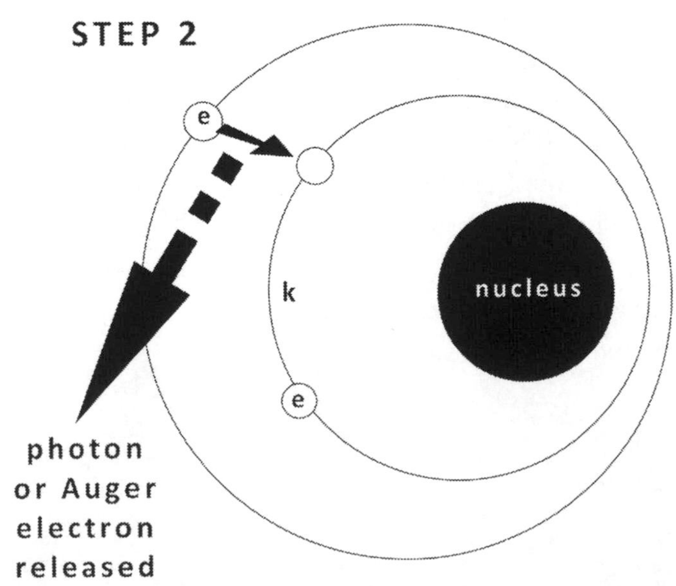
STEP 2
e
k
nucleus
e
photon
or Auger
electron
released

Another reaction, the Compton effect, occurs when a photon interacts with the electron of an outer shell. To understand this process, we must first review the electromagnetic spectrum. Recall, all electromagnetic waves travel at the speed of light in a vacuum (3×10^8 m/s) and are denoted as the constant *c*.

Note: c = frequency × wavelength for all electromagnetic waves.

Also note that the spectrum is characterized by the following:

wave type	wavelength meters	frequency hertz
radio	10^{3}	10^{8}
microwave	10^{-2}	10^{10}
infrared	10^{-5}	10^{13}
visible	10^{-6}	10^{15}
UV	10^{-8}	10^{16}
X ray	10^{-10}	10^{18}
gamma ray	10^{-12}	10^{19}

Returning to the Compton effect, a photon hits an outer shell electron and is ejected. An energy transfer occurs as the ejected electron (known as the Compton electron) absorbs part of the incident photon's energy. This lowers the energy of the scattered photon causing it to travel with a greater wavelength.

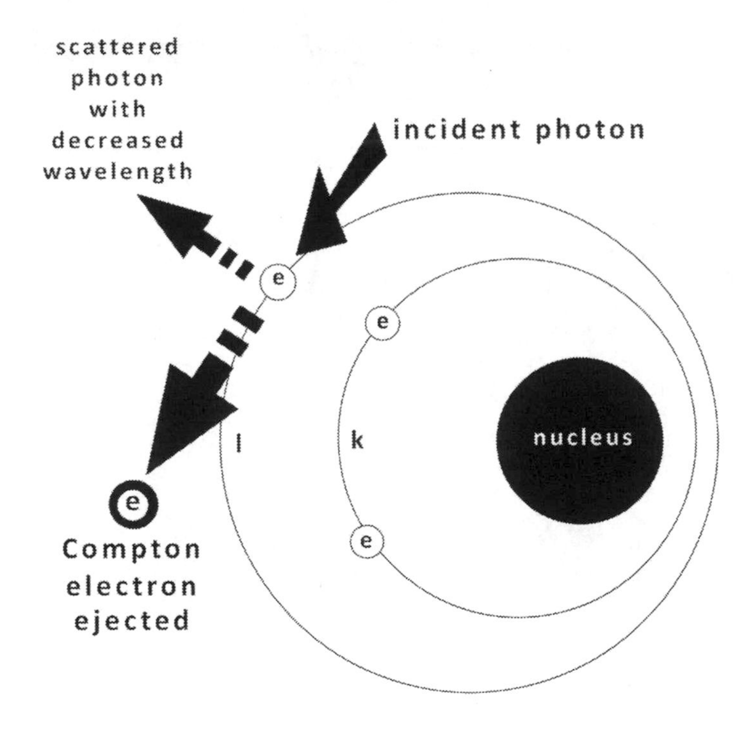

Finally, the process of photodisintegration occurs when a photon interacts directly with the nucleus of an atom. A photoneutron is released in the process, but the resulting composition of the nucleus may be an unstable product. This new atom may be radioactive and undergo further radiation emission to form a more stable product. The incident photon in this reaction must be of very high energy to cause such a reaction, usually greater than 8 MeV.

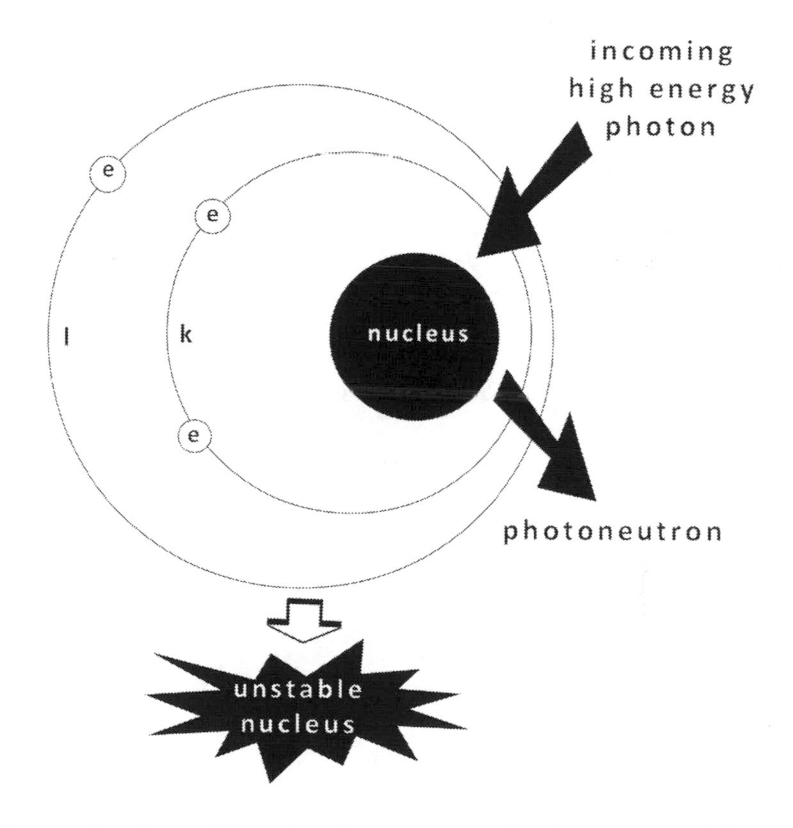

As previously mentioned, coherent scattering is not powerful enough to cause ionization of an atom. However, the photoelectric effect and the Compton effect both cause *direct* ionization with the removal of an electron. Photodisintegration involves interaction with the nucleus to provide *indirect* ionization of the atom.

Another interesting occurrence between the electron and atom results in the production of bremsstrahlung radiation. In this event, an electron is fired at the nucleus. No direct contact is ever made between particles, however, the electric field created by the atomic nucleus causes the speeding electron to "brake" and slow down. The decelerating electron releases energy in the form of an x-ray.

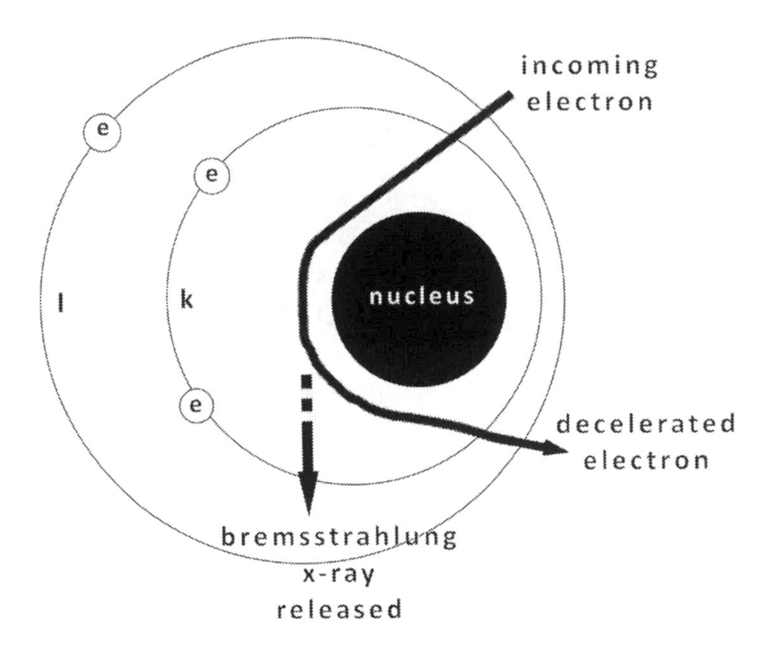

Heavy particles such as protons, neutrons, and alpha decay products may also undergo interaction with atoms. These collisions may be elastic (during which no energy is lost) or inelastic (during which energy is lost). Note that many of the electron reactions previously described are inelastic collisions.

In *neutron activation*, a neutron is absorbed into the nucleus and a proton is expelled. This does not change the atomic weight, but instead the atomic number. This may result in an unstable nucleus, similar to photodisintegration, causing nuclear breakdown and radiation release.

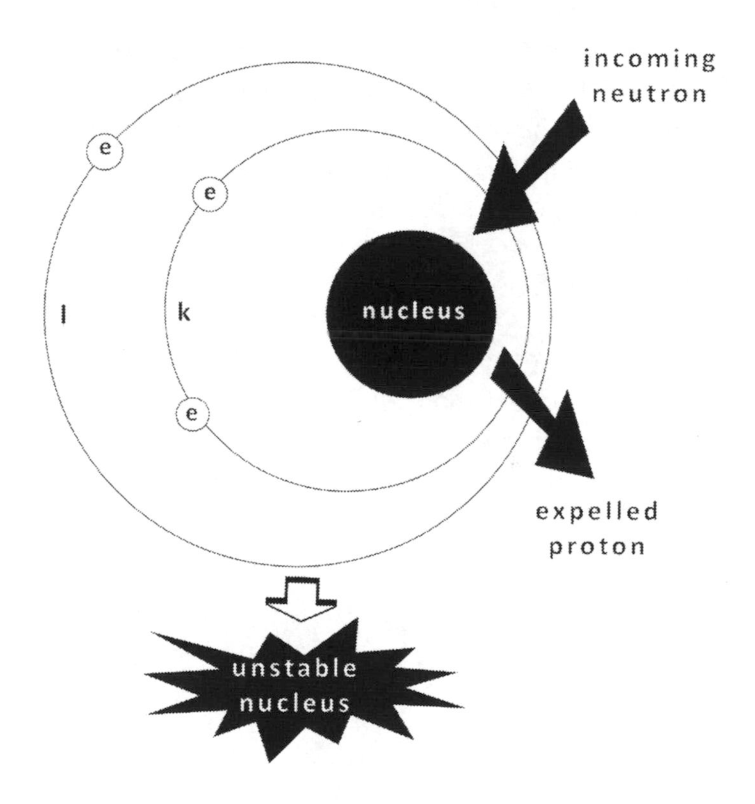

Dose

Now that we have a handle on how radiation is produced, we can relate this to how such a dose of radiation is characterized when applied to biological tissue. A dose, or *absorbed dose*, is the amount of radiation energy absorbed per unit of mass. The unit of absorbed dose, the gray (Gy), is equal to 1 joule per kilogram. The gray replaced the rad (**R**adiation **A**bsorbed **D**ose), a term used in older textbooks.

External Beam Dosing

The *percent depth dose* (PDD) is used to describe the difference between the level of radiation being delivered at two separate points and is expressed as a ratio. Consider an external radiation beam with the center of the beam being the point of greatest energy. This part of the beam makes contact with a certain material at point A and travels downward into the material to reach point B.

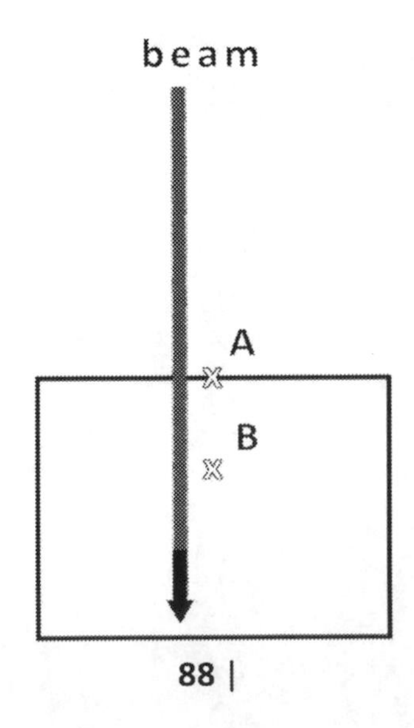

The surface dose is expected to be delivered at point A, and PDD compares this with the dose being delivered within the material at a certain point B.

Tissue-air ratio (TAR) seeks to characterize the differences between the doses being administered to the same area within a point of air and within a point of a certain material.

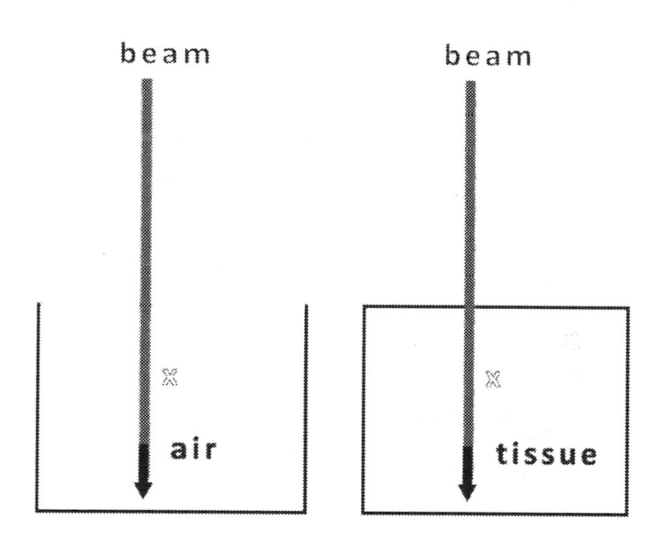

PDD is dependent on the source-to-surface distance (SSD), as quantified by the length from the machine to the material. However, TAR is independent of SSD.

Finally, the *tissue-phantom ratio* (TPR) compares doses delivered between a certain material and biological tissue. It is used for higher energy external beams (>4 MeV) when TAR is not useful. The points of reference are the same as those used for TAR, except that air is being substituted for a material substance.

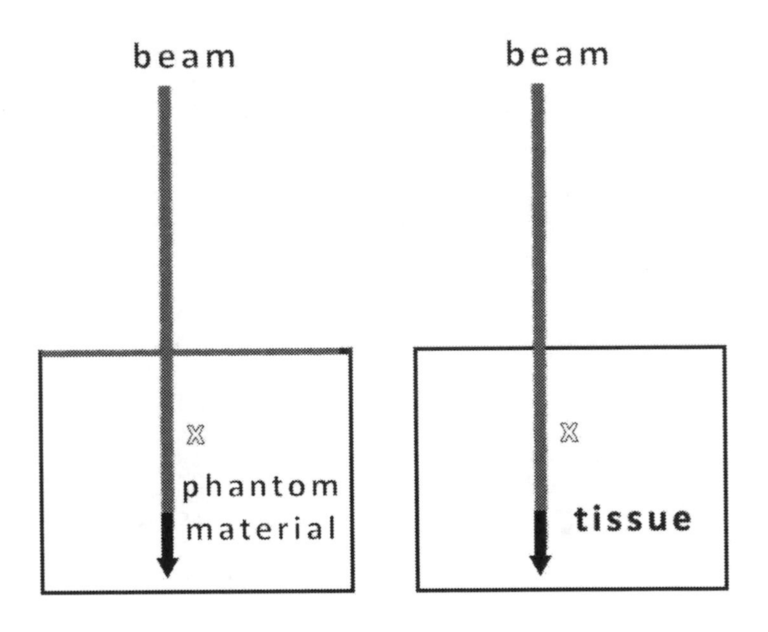

When calculating the radiation dose that a patient may receive, water is substituted as the phantom material, as this is the main component of tissue.

Much like the first few millimeters of a sponge soaking up water from a countertop, the first few millimeters of radiated tissue also soak up the photon beam. This is known as the *buildup region*. The dose within the tissue increasingly builds up to a maximum point at a specified depth (D_{MAX}) before tapering down.

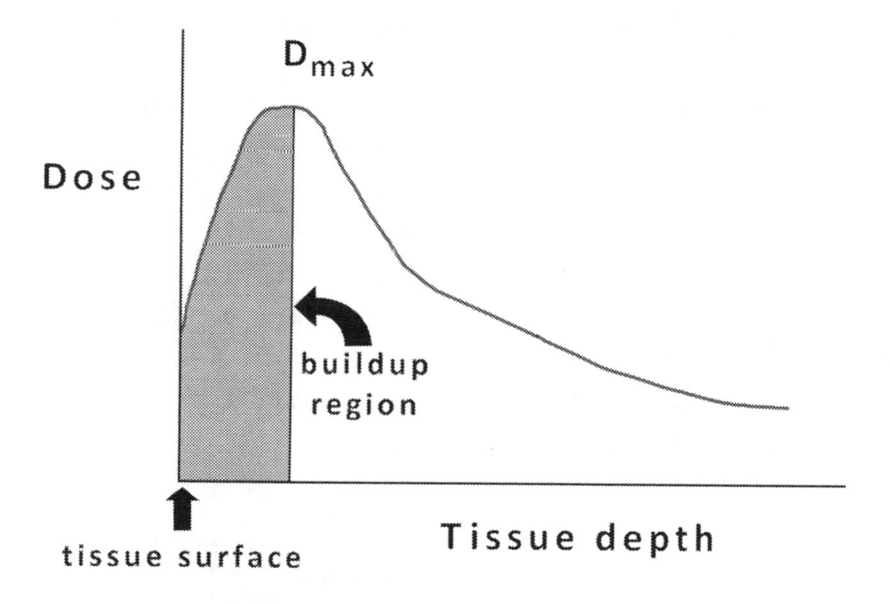

Therefore, the largest amount of radiation delivered is not to the surface of the skin, but in the tissue that may be millimeters or even centimeters deep.

Photon Beam Distribution

The radiation beam exiting the machine is not a perfect geometric shape that retains its form until it comes into contact with the patient. Instead, it is a slightly enlarging collection of many beams, some of which scatter. As later described in our radiation techniques section, external beam radiation exits the machine at the *gantry*. The gantry may have *trimmer bars* that assist in decreasing the scatter of the photon beam as it exits the machine.

The beam itself has various parts, with the greatest amount of energy (approximately 80%) being delivered between the *shoulders*. Less than 20% of the energy delivered will be in the scattered *toe* portion of the beam. Between the toe and shoulder is a small portion of the beam known as the *penumbra*. The penumbra is defined as the region from which the percentage of the dose delivered falls from 80% to 20%. A shorter penumbra indicates that a beam is sharper, whereas a wider penumbra (as in the case of Cobalt-60) indicates greater scattering. Trimmer bars therefore play a much more integral role with a Cobalt-60 machine. Also note that a higher energy beam causes greater scattering and a wider penumbra. The penumbra also increases with an increasing SSD and field size.

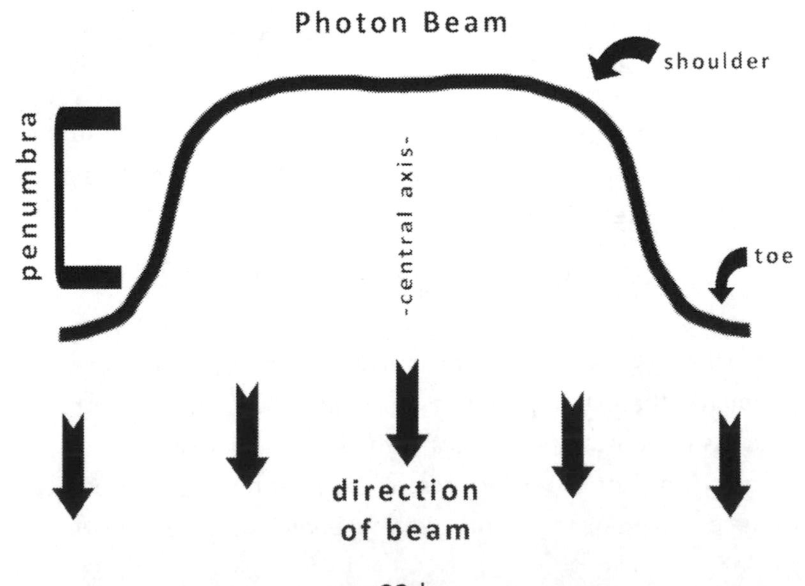

A 2-dimensional picture of the beam is considered above, but be mindful that photon beams are 3-dimensional in space. The source holds the photon beams as they exit, but they will continue to expand outward along all edges.

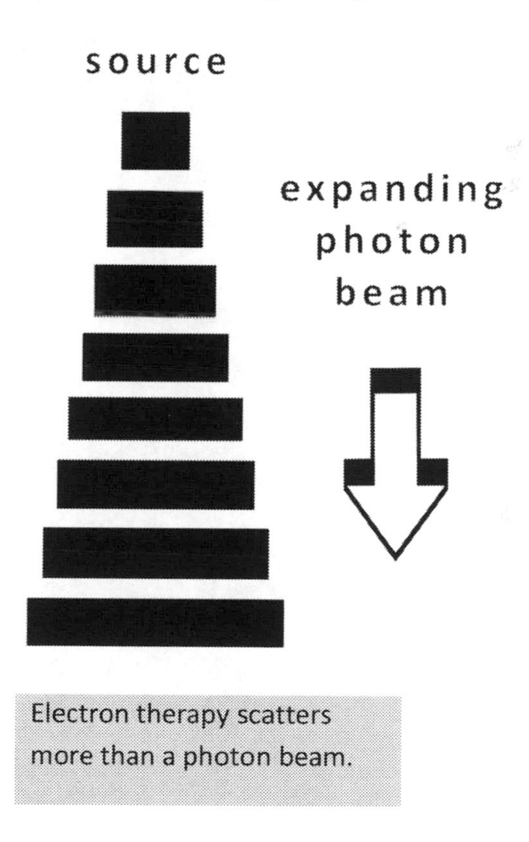

Electron therapy scatters more than a photon beam.

Additionally, at the source of the photon beam we can alter the distribution of the radiation with customized collimators, wedges, and blocks. This shapes the beam to a conformation that may allow for better patient treatment. This can increase the penetration of a beam to deliver a higher dose or merely prevent unnecessary treatment to otherwise healthy tissue.

SECTION THREE:

RADIATION TECHNIQUES

The changes in the field over the last 20 years can only be described as staggering. With improved technology we hope to more safely deliver higher doses to tumors while delivering lowered or perhaps stable doses to the normal tissues.

It is first important to differentiate teletherapy and brachytherapy techniques. "Tele-" is Latin for distance, although it is more commonly referred to as external beam radiation. Cobalt machines and linear accelerators deliver therapy from approximately 100 cm. Brachytherapy (translation: "near") delivers treatment at a close proximity, often internally with probes or seeds. For example, prostate cancer may be treated with Iodine or Palladium seeds (approximately the size of a grain of rice) that are implanted into the prostate via needle placement.

Our discussion will begin with the process the patient undergoes prior to external beam radiation and work through the initial consultation, simulation, treatment planning, plan evaluation, and delivery. A look at other common procedures used in radiation oncology will follow.

Radiation Delivery in a Nutshell

When first referring a patient to a radiation oncologist, they may need further radiographic imaging to localize the region of the tumor or to seek out other areas of tumor spread (i.e. nodal involvement) that can also be treated. Some physicians can do this in their office, whereas others may need more specific studies, such as a PET scan or MRI.

Once a treatment plan appears evident, a simulation of the radiation delivery is performed with the patient specifically positioned as they would be during the actual treatment. This simulation is merely a rehearsal of future radiotherapy sessions and can take anywhere from 20 to 60 minutes. Once organized, the patient lies on the treatment table and is set up as he/she was initially simulated. Once the patient is appropriately positioned (with the use of room lasers and external tattoo marks that were placed on the patient during CT simulation), *external portal images* (EPI) are taken. These are radiographic images of the area that is to be treated after positioning the head of the radiotherapy machine with any blocks in place. These images are reviewed by the resident and attending prior to treatment to verify that no positional changes are needed to appropriately deliver the radiotherapy treatment. To perform this, the CT images from initial simulation are reconstructed into images of the planned beam's eye view called digitally reconstructed radiographs (DRRs). If the EPI match with the DRRs (i.e., bony landmarks and blocks are in appropriate placement), then radiotherapy may begin. This is known as *block verification*.

Fractionation

The goal of fractionation is to take biologic advantage of the differences between the cancer cell and the normal cell. Generally, cancer cells are faster growing than normal cells. However, the mucosa of the aerodigestive tract itself is rapidly overturning, and thus is relatively similar to the rate of tumor growth.

Standard fractionation is typically delivered as a dose of 1.8 Gy of radiation per treatment session once a day for five days a week. Weekend days are usually spent as days off, unless the treatment regimen calls for it. An individual treatment is painless and lasts only a few minutes.

Because each patient is different, the use of various fractionation schedules may be appropriate. For patients with aggressive tumors that will repopulate quickly, twice-a-day treatments may be indicated. *Hyperfractionation* is referred to as a treatment regimen that gives less than standard doses (<1.8 Gy), no less than 6 hours apart, to deliver a higher total dose over a planned treatment course. The hope is that while acute toxicity may be worse, long term toxicity is decreased and thus a higher total dose is tolerable. Randomized studies in the head and neck have supported the concept.

The method of delivering multiple standard doses of radiation per day to deliver a total radiation dose in a shorter time is referred to as *accelerated fractionation*. This decreases the amount of time between treatments, limiting the amount of tumor cell repopulation that occurs. This also decreases overall treatment time.

In cases of oncologic emergencies, very large doses of radiation are administered in an effort to assist in reducing the size of tumor masses that may be impinging on the spinal cord or obstructing the airway. These doses may be a one-time dose or fewer in number given their potential for toxicity. Radiation delivery is discussed in more detail later when we review radiation techniques.

Radiosurgery uses physics and radiobiology to accurately deliver a very high dose to a very sharply delimitated target volume. Radiosurgery is widely used for brain metastases and has produced very encouraging local control rates for small volume, localized non-small cell lung cancer.

External Beam Radiation

Initial Consultation

It is important to understand the expectations of the various team members within a cancer institute and the radiation oncology department. These professionals work directly with the physician to ensure the best possible treatment for patients.

Medical Dosimetrists

Dosimetrists are board certified professionals that work with patients and computer models to help deliver the prescribed radiation dose as written by the radiation oncologist. They must take into account the area being radiated to help avoid complications to surrounding organs and important anatomical structures. Some tissue, such as nephrons and nerves, is dose-limiting and its amount of exposure to radiation must be minimized.

Physicists

Medical radiation physicists supervise the efforts of the dosimetrist and assist them on complex treatments plans. Quality assurance is another big responsibility, as the beams of the various machines must be routinely tested to ensure precision. They are also a board certified profession, and many assist in the educational department constructing the physics curriculum for residents at training facilities.

Radiation Therapist

Certified by the American Registry of Radiologic Technologists, these team members administer daily radiation therapy. They make records of the patients' visits and make sure the fields are in place at every visit. They also help make sure the machines are working properly and check them routinely.

Surgical Oncologists

Although not part of the radiation oncology department, a good surgical oncologist is able to assist in patient care by performing an initial biopsy. Often the biopsy is taken from an area of convenience with minimal intervention. If a biopsy or exploratory surgery is required, the surgeon will be able to help stage the patient by searching for and documenting the extent of tumor spread. Commonly, the surgeon resects only primary tumors that have not spread, or in some cases, a solitary metastasis. During resection of macroscopic disease, titanium clips are placed at the site of the malignancy. This assists the radiation oncologist who will later target the area with radiotherapy for potential microscopic spread.

> Because wound healing is impaired by radiation therapy, surgical resection of a tumor requiring the re-anastomosis of a structure (example: colon) may only be irradiated on one side. This allows the other portion of the newly joined structure to promote healing.

Because of the prevalence of malnutrition in patients suffering from cancer cachexia, the surgical oncologist or interventional radiologist may also help by placing a PEG tube to promote enteral nutrition.

Radiologist

Appropriate staging with a review of the radiographic studies is important in determining what areas may require treatment. An experienced radiologist familiar with patterns of cancer spread can be a vital asset.

Pathologist

Understanding the histologic diagnosis is important prior to radiation because a biopsy of previously radiated tissue alters its cellular appearance. The approach to squamous cell carcinoma versus adenocarcinoma can mean different surgical, radiation, or chemotherapeutic treatments.

Medical Oncologist

Referrals for radiation therapy are typically initiated by the medical oncologist. Treatment with chemotherapy may need to be initiated before, during, or after the course of radiotherapy. The multidisciplinary approach to treating cancer is the focus of a tumor board which allows the medical oncologist to discuss treatment options with the surgical subspecialists, pathologists, radiologists, and the radiation oncology department in a manner that provides a consensus among all caregivers.

Ancillary staff

Many other people help comprise a good cancer institute. Nurses specializing in administering chemotherapy have experience monitoring patients for side effects. Registered dietitians are often available for consultation to help avoid malnutrition with central parental nutrition and PEG feeding recommendations when necessary. If patients are undergoing radiation to the head and neck area, it is not uncommon for temporary swallow difficulties to require the assistance of a qualified speech therapist.

Simulation

Before external beam and brachytherapy treatment, many factors must be considered to avoid any unnecessary harm and to maximize the potential of therapy. This takes place among a number of professionals within the radiation oncology department. Some aspects of the work can be completed on paper, and other considerations are discussed among the physicists, dosimetrists, technicians, and the physician. Most of this work is completed while the patient is present in a situation referred to as *simulation*.

Simulators are rooms with machines setup to mimic the actual treatment room. The simulator has the ability to use radiographic and fluoroscopic images to analyze the area requiring treatment. This assists in forming a specified plan by which radiation can be delivered. Simulation for a given patient may take approximately one hour, but this will vary depending on the complexity of the malignancy. After simulation, however, actual treatment may last no longer than a few minutes. In addition, because no actual treatment occurs in simulation, there is no competition for the room between patients requiring treatment and patients undergoing preparation for treatment.

> Prior to simulation, the process of *clinical setup* was widely used. In clinical setup, the patient was taken to a treatment room and radiation fields were chosen based on the best possible previous radiographic evidence of malignant disease, alongside thorough clinical examination. A port film, a non-diagnostic x-ray, was taken to evaluate the beams while the patient is in position. The port film is hazy, difficult to read, and appears crude in comparison with today's technology.

Since multiple beams are used for almost all treatments, the linear accelerators rotate about a point in space known as the isocenter. Usually the isocenter is placed within the center of the tumor and the gantry swings around the tumor to reach the position to deliver appropriate radiation. All conventional linear accelerators use a source-to-axis distance (SAD) of 100 cms. The Source to Skin Distance (SSD) technique is used for some simple, single field setups such as a spinal cord. SSD uses a specified distance between the patient and the source, typically 100 cm. Each time the patient presents for treatment this distance is measured and used as a reference point. Because multiple treatment fields are required, the patient must be moved between each field and the distance is measured to 100 cm each time. Because SSD is very time consuming, SAD is now much more commonly used.

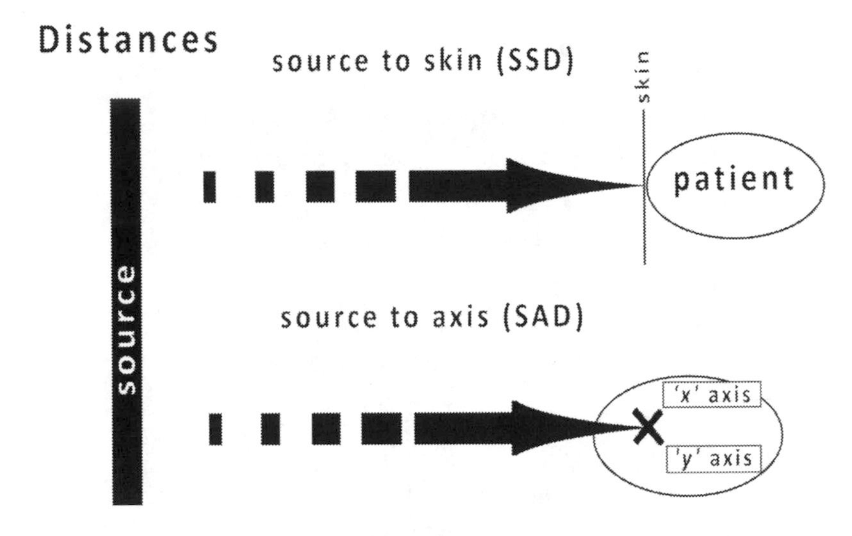

In either simulation process the patient is placed in the treatment position, usually lying supine on the table. The table itself is rotational in all directions so that the best possible treatment fields are attainable. Previous medical information on the patient in addition to real time fluoroscopic or radiographic data is used to find the SSD or SAD. Important anatomical structures like nerve bundles are also considered when constructing a field for treatment. Avoidance of these areas, if possible, will lead to fewer complications.

To allow for greater precision of dose delivery, a patient must be able to maintain the same position during each fraction. Patient comfort is a priority, as this creates better stability, and the patient is less likely to move. The patient may be sitting, standing, prone, or supine. The supine position on the table is the most common, and machines are made to accommodate this assumption. However, some thought has been given to comparing patient positions for those being treated with prostate cancer, and no one way is standardized.

Immobilization structures may be created for the body part being irradiated. For head and neck cancer treatments a mesh facial mask is custom fitted for the patient, and it attaches directly to the table to maintain stability. When final field arrangements are made, the patient may be tattooed at the site to mark the area of therapy. This helps to position the treatment beams on the patient in the same place on every visit.

Given recent advancements in radiographic imaging, it is now possible to perform *virtual simulation*. This setup draws information about the patient's anatomy from a CT scan and uses it to form a 3-dimensional picture that will determine the geometry of the treatment area. From there, dosimetrists can evaluate the data to calculate the necessary beam energy.

Treatment Planning

Clinical setup, although popular during the 1950's and 60's, is no longer practiced in modern medical facilities. This treatment consisted of rectangular shaped external beams that needed to be altered based on the patients disease process. The beam could only be customized by blocks that were created for each patient, the shape of which was derived with the help of x-ray films and fluoroscopy.

With recent advancements in computer technology, the field of conformal radiotherapy is able to create a more detailed 3-dimensional treatment volume as opposed to previous 2-dimensional volumes. This new approach is referred to as 3D-CRT.

2-D application of treatment beams

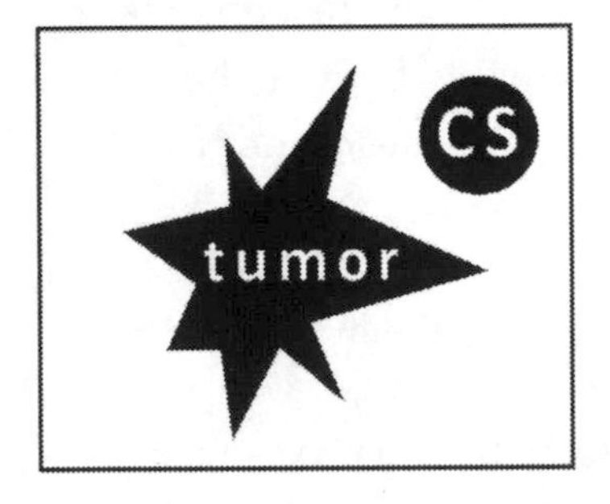

3-D conformal treatment area

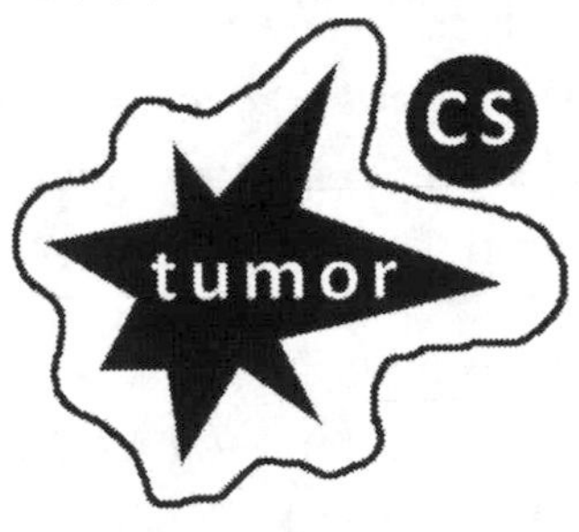

If we consider “CS” to be a critical structure in the picture above, the 2-D treatment beams will cause radiation damage. The volume of the area treated in the 3D-CRT is able to omit or at least minimize any delivery of radiation to CS. The advantages of this technique further the sparing of healthy tissue and critical anatomical structures.

Conformal Therapy, with the help of the CT scan, has revolutionized the field of radiation oncology. Introduced in the 1970’s, physicians were able to visualize a cross section of the tumor for the first time. Being able to view slice by slice details of the patient then allowed for a 3-D image of the tumor burden to be conceptualized. Through advances in computer software programs, a collaboration of MRI, PET scan, and CT imaging allows for the best possible assessment of the anatomy with an emphasis on creating a volume to which the dose of radiation can be delivered. With several beams directed at various angles, as well as the modification of these beams by multileaf collimators (discussed later), various doses can be delivered at a time.

While the CT scan is important in understanding the structure of a tumor, the MRI depicts soft tissue better. A PET scan may highlight smaller, more remote involvement. Taking all of these imaging studies and combining them to help identify a tumor's geometry is known as *image fusion*. Using multiple studies is also helpful in *contouring* a lesion, however it is the data set from the original CT scan that is most referenced. Repeat studies attempt to match the original CT scan slice for slice to reassess and treat.

The contouring process involves detailing the anatomical structure on every CT picture which may include slices as close as 1 mm. Artifact is often seen in radiographic studies, and a best judgment must be made to determine the true contours of the tumor. In addition to critical structures that require identification, one must also recognize that certain organs are limited in the amount of radiation they can receive. Therefore, it is also important to contour surrounding organs to assess for potential dose delivery to these regions.

Treatment areas

Deciding on the volume of the treatment goes beyond what may be visible on radiographs. The various volumes of tumor therapy are:

Gross tumor volume (GTV) – The volume containing the actual tumor as visualized on exam or on radiographs.

Clinical target volume (CTV) – The GTV plus the surrounding area which accounts for any microscopic disease that is assumed to be present. This margin may extend the GTV by 5 to 10 mm, and should receive the prescribed dose to be delivered. Margins of one plane may vary depending on the characteristic spread of the malignancy. For example, if a tumor is typically shown to extend downwards, the inferior margin may be extended to include 10 mm of tissue, whereas the rest of the tumor may only require a margin of 5 mm.

Planning target volume (PTV) – When a patient is on the table, many actions are taken to prevent movement. This allows for the same positioning of the patient during each treatment so that precision can be maximized. A few millimeters may change despite the best efforts. Making adjustments in treatment planning for shifting pelvic structures and movement with breathing are part of *4-D conformal therapy*. Accounting for these shifts requires changes in the calculations of dose and beam configurations. The volume radiated within the PTV is intended to be large enough to account for the CTV and any movement.

Treated volume (TV) – Ideally, TV and PTV would be the same, but healthy tissue without microscopic evidence of disease may be included in the outer boundary of this volume.

Irradiated Volume (IV) – With scattering of photons and body movement, the total irradiated volume accounts for the area receiving a significant amount of the radiation delivered. Dose-limiting organs and important anatomical structures must be accounted for and avoided if possible when computing this volume.

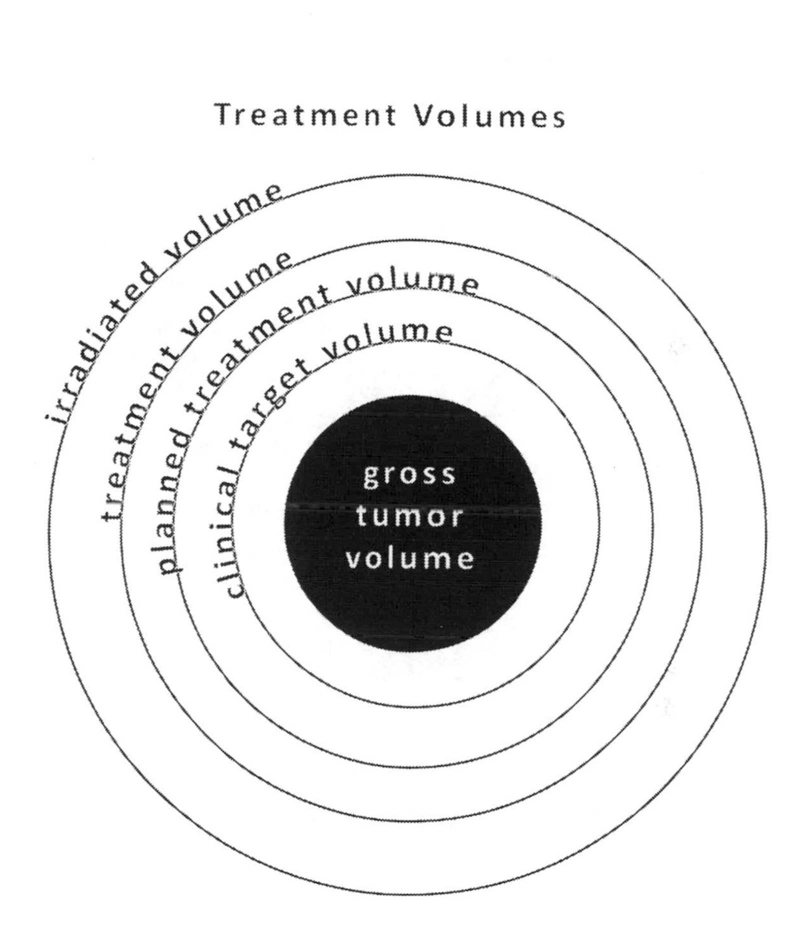

For documentation purposes, it is recommended that every patient have a *reference point* within the planned tumor volume. This reference point serves as a location to which the dose delivered can be accurately calculated. This also means the area must be easy to identify and is usually chosen along the line of the central axis of the external beam or within a dense portion of the malignant tissue.

Beam modifying devices

The radiation delivered by machines can be altered in many ways. A few of the techniques used are described below.

Blocks and *multileaf collimators* (MLCs) are devices used to assist in forming a more detailed geometric shape suited for the intended treatment field. The beam exiting the machine is typically a rectangle, but the full use of that beam within the rectangle may not be necessary for the patient. Blocks are blunt shapes that may be fitted to the machine at the origin of the beam.

rectangular treatment beam

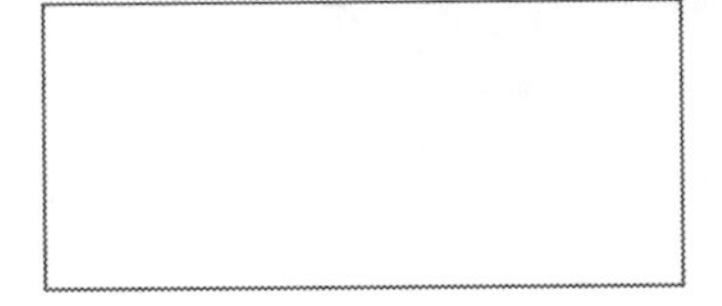

beam with triangle block

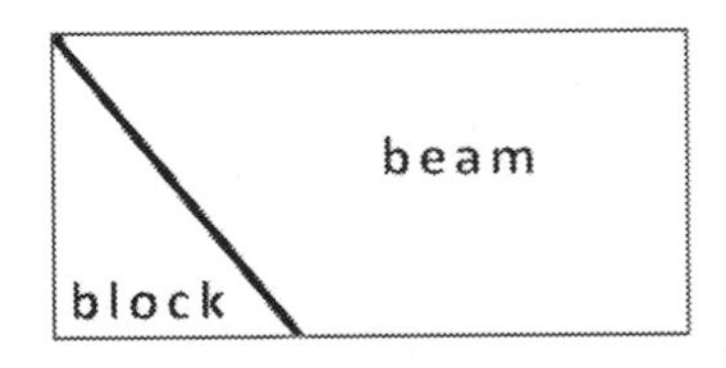

beam with multileaf collimator

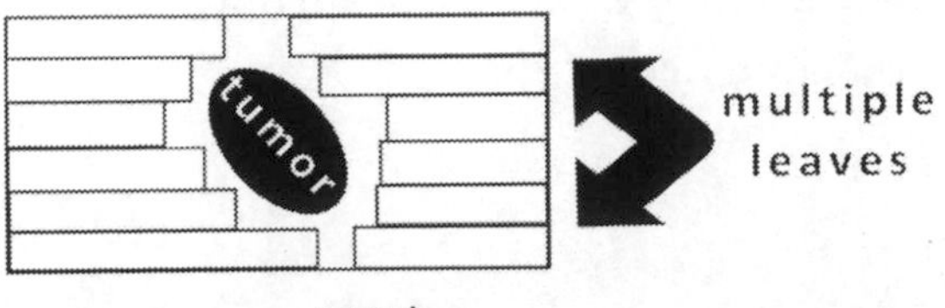

Multileaf collimators are unique because they are built directly into the head of the machine. Each leaf is quickly altered to produce a specific shape thereby changing the beam for each treatment field. This decreases total treatment time because there is no need to switch out blocks between fields. MLCs and blocks must be made from heavy metals like tungsten and lead to absorb the radiation beam. Whether using a block or MLC, the conformation of the beam allows a customized treatment for each patient.

Wedges alter the beam conformation and typically take a triangular shape. Wedges are usually placed on the machine at the origin of the beam like blocks. However, they do not block radiation therapy. Wedges act to help evenly distribute radiation when the contour of the body is sloping or the thickness of the skin is varied. Should no wedge be present, a thinner portion of the skin would allow for a greater amount of radiation to pass through, whereas a thicker portion of the skin might not receive as much. *Tissue compensators* work in a similar manner when placed onto skin with an irregular surface.

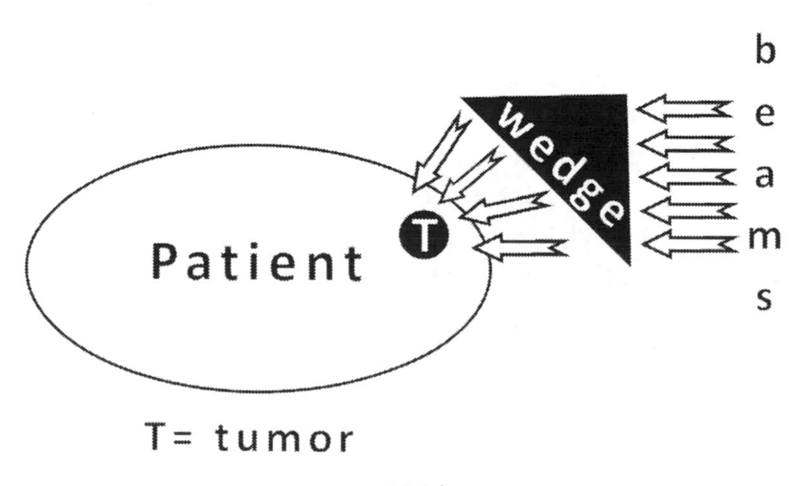

Lastly, recall from physics that the maximum absorbed dose is at a depth D_O and not along the skin surface. The higher energy beams are relatively skin sparing, and it may be necessary to add a *bolus* to the top of the skin when undergoing radiotherapy. The bolus, also called a *tissue equivalent,* takes on the characteristics of the skin and absorbs the first few millimeters of radiation. This allows for the D_O to include a heavier dose to the skin itself instead of being spared. This would be important, for instance, in patients with inflammatory breast cancer given the concern that malignant cells may have migrated to the skin.

Forward and Inverse planning

Two types of planning models can be used: forward and inverse. Both accomplish the same thing, as there is a defined dose that must be delivered to the tumor site while limiting the amount of radiation exposure to surrounding structures.

Initially, all radiotherapy was developed with forward planning. The anatomy is defined and target volume is first identified. Beam arrangements and the addition of blocks or wedges further refine the area involved. From the beam arrangements chosen, a calculation of the dose being delivered can be quantified. An evaluation of the dose distribution is then scrutinized by the planning team. Modifications are made to improve the plan until it is finally accepted and implemented.

While forward planning determines what various dose distributions may be achieved, inverse planning chooses a dose up front. Working backwards, computerized models can then adjust the beam arrangements to fit the requests of the planning team. Various approaches may be available, so certain tissues are prioritized to receive less radiation. It is more important to spare the spinal cord than the lung, and computer software is able to take this into account when developing a plan for the patient. The computer is then able to determine beam shape, angle, intensity, and weighting.

The best example to differentiate forward and inverse planning is that of a fancy toaster*:

Forward planning - I decide on the temperature and time, and then check to see if I under- or over-cooked the bread.

Inverse planning - I set the toaster to "golden brown," and the toaster decides the temperature and time.

*please shy away from all cooking analogies in radiation oncology when possible

After much consideration, a prescription for radiation therapy is written by the radiation oncologist and includes:

1. General information: name, date, medical record number
2. Site of treatment
3. Location of intended dose
4. Number of fields
5. Field arrangements

6. Energy
7. Total dose
8. Number of fractions
9. Dose per fraction
10. Frequency of fractions expressed as number of treatment/days per week. Example: a person receiving 2 treatments per day for 5 days a week would be written as 2/5.

Plan Evaluation

Once a plan is created, a thorough assessment follows. The plan can be rejected, modified (known as *iteration*) or accepted. Most plans are modified prior to being implemented.

Individual cuts in both the axial and coronal direction are evaluated for dose delivery to the tumor and surrounding organs. Newer computer models will use a multicolor diagram to estimate the value of the dose administered at each pinpoint location. The dose administered to the spinal cord on a specified plan may be shown to be 60 Gy, an unacceptably high amount. Protocols for conformal therapy limit the maximum amount of radiation that individual organs may receive, and this sample plan would be rejected.

The importance of the isocenter comes into play when configuring the dose distribution throughout the irradiated tissue. The intended dose to be delivered is aimed at the isocenter and referred to as the *plan normalization point*. The PTV should receive at least 95% of the dose that is being delivered to the isocenter. This ensures that the entire volume of the tumor is being adequately treated. It is also important to have an equal *distribution* of the dose delivered throughout the tumor volume. A maximum dose delivered within the PTV may be set at 5 to 10% greater than that received by the isocenter. If a plan has too much variation in the dose delivered throughout the PTV or too little dose delivered in one portion of the PTV then it is likely to be rejected.

Patient toxicity: LD_{50}, TCD_{50}, $TD_{5/5}$, & $TD_{50/5}$

Toxicity to normal tissue has been thoroughly studied. Under- or overdosing a patient can be critical, and a few definitions need understanding. Too great a dose creates toxicity, and too small a dose will mean ineffective treatment for the patient. Failure on either end will mean rejection of the plan.

An **LD_{50} assay** is the radiation dose expected to produce lethality in 50% of subjects, as derived from a dose response curve.

TCD_{50} is the tumor control dose, the total radiation dose required to achieve local tumor control over a specified period of time in 50% of patients.

$TD_{5/5}$ is the dose of radiation expected to cause complications in no more than 5% of the patients treated within 5 years. This dose is considered to be an acceptable risk for patients undergoing radiotherapy. It is otherwise known as the minimal tolerance dose (MTD) and varies depending on the tissue that is targeted with treatment.

$TD_{50/5}$ is the dose of radiation in which half of patients are expected to develop toxicity within the target organ within 5 years.

On a much smaller scale, **D_0** is the dose required to produce one lethal lesion in a radiated cell. For mammalian cells, this is commonly 1 to 2 Gy.

Delivery

External beam therapy or *teletherapy* may be delivered by machines containing radioactive material (such as cobalt) or linear accelerators (a.k.a. linacs). Both have a similar structural appearance, though the components inside are very different. Note that a gantry crane is a large overhead hoisting crane, typically seen at ports transferring large cargo on and off ships. The gantry portion of the teletherapy machine takes this name because it operates so similarly. The gantry has rotational ability to move about the patient overhead and side to side while delivering radiation. The gantry is connected to a stabilized gantry island, allowing the treatment head to rotate around an axis.

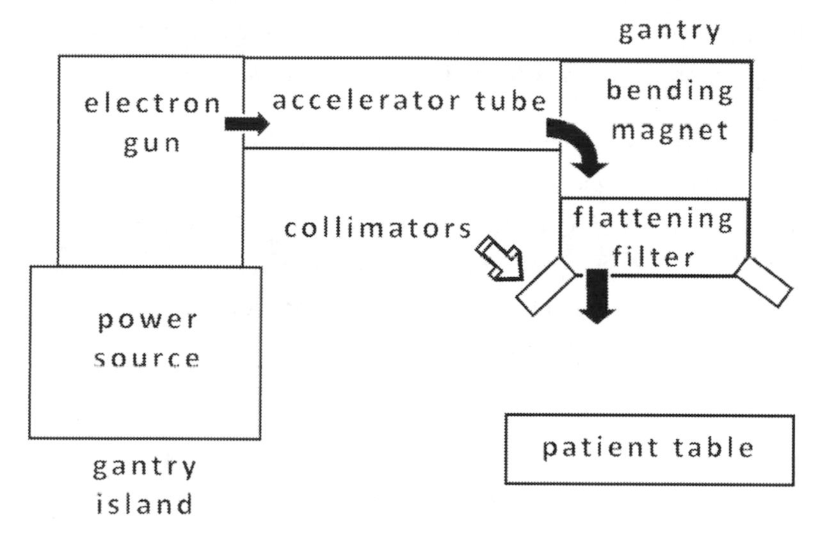

LINAC OVERVIEW

For many years, radiation treatment occurred as a result of exposing patients to the controlled radioactive decay of certain species such as Radium-226, Cesium-137, or Cobalt-60. Cobalt machines were the standard of radiotherapy for many years until linear accelerators gradually replaced them. The radioactive decay of the unstable Cobalt-60 species emits two gamma rays and is transformed to Nickel-60 in the process. The machine itself holds the radioactive material in a compartment of the gantry overhead that is opened only during treatment or calibration. A physicist in the radiation oncology department is responsible for insuring that the proper amount of radiation is being released from the radioactive material. In the case of the Cobalt-60 machine, approximately 200 cGy should be released per minute at a distance of 100cm from the patient. Collimators (also shown on the linac above) are devices used to filter and direct the radiation emitted from the machine, thus assisting in the shape of the area to be radiated. Further definition with trimmer bars on the gantry prevent the escape of any extraneous radiation.

Cobalt-60 carries a half-life of 5.3 years, and the dose being administered gradually decreases over time. The need to replace the radioactive Cobalt in these machines occurs approximately every 6 years. Although these machines are becoming a thing of the past, some are still being used with effectiveness in head and neck cancers. Like all external beam therapies, treatments take only a few minutes and are painless for the patient.

Unlike the cobalt machine, linacs use electricity, not radioactive elements, to produce photons. The bremsstrahlung effect , a phenomena first observed over a hundred years ago by Nicola Tesla, is used to produce radiation with linacs. To start, a power source to the electron gun provides the initial electrons that travel down the accelerator tube with the help of a magnetron or a klystron that is producing microwaves. The accelerator tube connects to the treatment head of the gantry before making a 90 or 270 degree turn downward toward the patient table. The electrons change direction as a result of a bending magnet at the end of the accelerator tube. Now poised toward the patient, the electrons hit a transmission target producing x-rays. A flattening filter then ensures that the radiation across the field is pointed in a uniform direction. Another device, the dose monitoring filter, consists of sealed ion chambers that monitor the dose rate and field symmetry. Finally, collimators adjust for the specific shape of the treatment area desired.

There are two different types of linacs: stationary wave and traveling wave. A stationary (or standing) wave created by a linac maintains a constant position, and its reflection travels along the wave's initial path. A traveling wave may fluctuate, as seen the in following depiction.

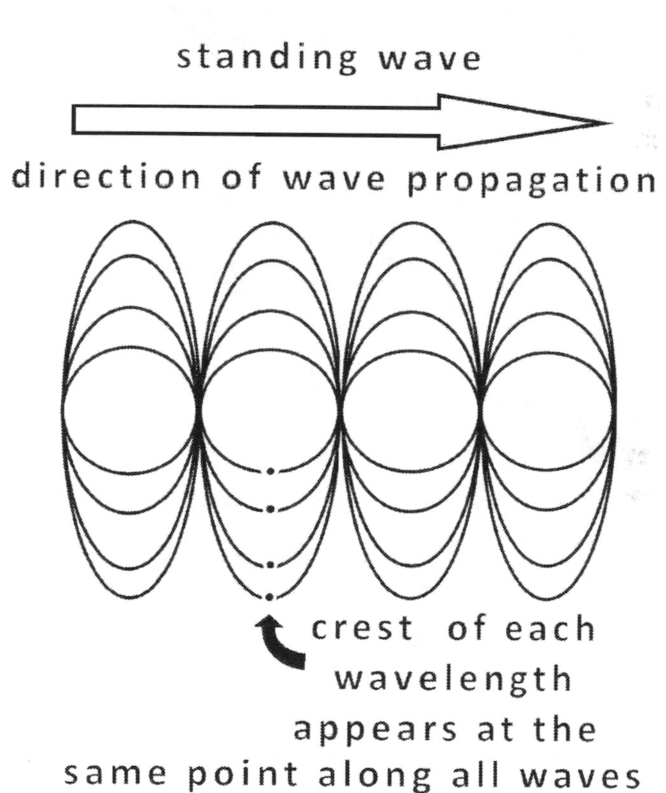

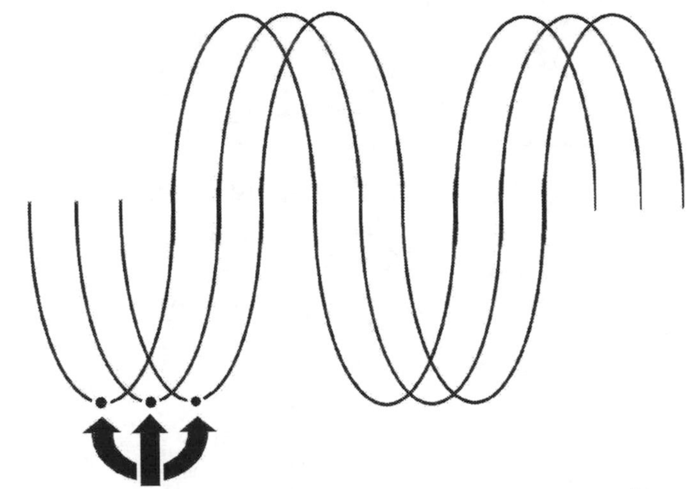

The world's largest particle accelerator at Stanford University is 2.0 miles long. Operational since 1966, it is buried 30 feet below ground and used for physics research. It may accelerate electrons and positrons up to 50 GeV.

Intensity modulated radiotherapy

Also referred to as IMRT, this technique takes 3D-CRT and improves upon it by allowing for the beams to deliver varying doses within the target. This allows for better sparing of normal tissue with a tighter conforming treatment volume. Because of the complexity of the structures surrounding head and neck and prostate cancer, use of IMRT has become increasing useful for treating these malignancies. Various methods of administering IMRT include Tomotherapy® IMRT, static IMRT, and intensity-modulated arc therapy (IMAT). Inverse planning is used with IMRT techniques.

Tomotherapy ® IMRT uses a narrow slit beam similar to that of a CT scan to deliver a dose at a time. The linac delivering the radiation has a MLC that changes throughout treatment as the beams are administered in a line by line method.

In static IMRT, the gantry rotates around the patient in an arc formation allowing several angles at which it may deliver radiation. At each angle the configuration of the MLC may be changed several times to deliver different amounts of radiation to the respective angle.

IMAT is able to deliver radiation at angles along the arc like IMRT, but with continual dosing between angles as the gantry swings around the patient. This extra radiation is referred to as a "superimposing arc". The radiation given during the arc is factored into the planned treatment dose and the total amount of the intended dose for the patient is not exceeded.

Volumetric arc therapy has improved on IMRT by being able to deliver a more precise 3-dimensional treatment volume in a shorter time period. In one 360-degree rotation of the linac arc, the entire planned treatment volume is delivered in a manner that would otherwise require several slices of IMRT.

Image-guided radiation therapy (IGRT)

IGRT uses fluoroscopic and radiographic images to monitor the target volume during treatment, and specialized software varies the beams based on patient position. This allows for the most accurate delivery of radiation, keeping the TV/PTV value close to 1. With minor movement, respiration, or change in tumor size over the course of treatment, IGRT is able to adjust. Programs quickly turn beams on and off depending on the patients' position. If a patient moves and the tumor volume is shifted, the beam is turned off. The tumor is only radiated when it is seen to be in the proper position. This therapy highlights the use of 4-D technology in radiation therapy.

Stereotactic Radiosurgery

Radiosurgery is the practice of using a single treatment of high dose radiation on tumors with a volume less than 20 cm^3. In a single fraction, the technique is able to deliver upwards of 12 Gy, whereas conventional radiotherapy may only deliver 2 Gy per fraction. Both Gamma Knife® and linacs are able to perform stereotactic radiosurgery. When comparing multiple fractionated treatments of conventional radiotherapy to one session of radiosurgery, the latter is preferred. Tumor growth between fractions of conventional radiotherapy may make the tumor more difficult to control, and the exposure of healthy tissue to excess radiation is minimized with radiosurgery.

Gamma knife®

The practice of delivering such a high dose of radiation at one time without significant complications comes from the use of multiple beams. A typical cobalt-60 Gamma Knife® machine has 201 different beams directing radiation at a single focus. In this scenario, a patient is fitted with a helmet containing 201 holes that allow for the radiation to pass through. The patient is then placed on the table, and the helmet is attached to the machine to assure that no movement takes place during the procedure. Different helmets are available to vary the size of the beam from 4 to 18 mm in diameter. Imaging studies then locate the tumor and analyze the volume, computing coordinates for therapy. Each of the 201 beams has an individual collimator to further refine the geometry of the treatment volume. The Gamma Knife® is able to deliver 300 cGy per minute, and the procedure takes approximately four hours.

Linear accelerator for stereotactic radiotherapy

Linac radiosurgery works similarly to the Gamma Knife® in many ways, requiring immobilization of the head to maintain stability within the radiated field. Source to isocenter distance remains at 100cm, as with traditional linac radiotherapy. Collimators are able to adjust beam diameters between 4 and 40 mm as they follow along an arc around the patient. Like the Gamma Knife®, a neurosurgeon or radiation oncologist can image the lesion, develop a treatment plan, and perform the procedure all in one day.

Brachytherapy

As described at the beginning of this section, brachytherapy delivers radiation within close proximity to the tumor. This can be on the surface of the skin or inside the body via an applicator. The radiation dose delivered from such a short distance is much more potent than that from 100 cm as administered by external beams. The main principle that allows radiation oncologists to use this technique is the rapid decrease in dose when administering therapy at such close range. The dose supplied to tissue within a millimeter is very high, whereas the dose reaching a centimeter is near zero. This is different from external beam radiation which does not allow for such a steep gradient of dose distribution.

When physicians and physicists first took advantage of brachytherapy in the 1920's, radium-226 (a naturally occurring isotope) was used. Since that time the brachytherapy has been refined into high dose rate (HDR) and low dose rate (LDR) treatments. With it, we have applied newer isotopes to therapy. Irridium-192 is most commonly used for HDR, whereas palladium-103 and iodine-125 are used in LDR. Despite the differing rates by which brachytherapy can be administered, HDR and LDR may not differ in the total dose a patient receives.

In the 1950's it was evident that personnel in the radiology department were being exposed to these high doses of radiation along with patients. The practice of *afterloading* was developed to minimize the risks of exposure. Materials for LDR like I-125 and Pd-103 do not require a large amount of shielding, however, HDR treatment with Ir-192 requires a half value layer of 2.5 mm of lead. In HDR, the treatment is administered remotely after technicians have vacated the room. This is necessary as HDR delivers 2 to 3 Gy per minutes, similar to conventional external beam therapy, and must be performed under more controlled conditions. LDR has a much slower rate at 0.4 to 2 Gy per hour. For this reason, HDR is delivered in a fractionated treatment regimen, whereas LDR may be implanted in the patient on a temporary or even permanent basis.

Half value layer (HVL) is the amount of material required to decrease the exposure from a radioactive source to ½ of its unshielded value.

Brachytherapy may be given by three different means:

1. Radiotherapy to the skin may be applied by surface applicators for the treatment of skin cancer.
2. *Intracavitary brachytherapy* allows for delivery of radiation via preexisting body cavities.
3. *Interstitial brachytherapy* positions the source of the radiation directly into body tissue.

Special applicators were invented to safely administer brachytherapy, and simulation has made it possible to test several configurations prior to the exposure of radioactive material. Today *dummy* sources are placed in the treatment position and radiographic imaging can assist with positioning during simulation. Planning brachytherapy also includes choosing an isotope, determining an applicator, and calculating the insertion time. Afterloading by remote control keeps the patient in a designated treatment room and allows staff to administer therapy from a control panel outside.

Beyond cancer, endovascular brachytherapy has been used for the purpose of treating renal and coronary artery re-stenosis. However, with the development of drug-eluding stents this practice has been largely abandoned.

Superficial Radiotherapy

Superficial radiotherapy is a form of brachytherapy that uses an orthovoltage machine consisting of an x-ray tube that delivers anywhere from 50 to 500 kV of energy. Machines are mobile and used for lesions that are less than 3 cm deep.

First popularized in the 1950's, these machines were used to treat squamous and basal cell skin cancers, but were later phased out as Moh's surgery became more popular. However, in patients with primary skin cancers of the head and neck region, surgery can be difficult because of the sensitive structures in this region. In addition, some regions of the head and neck may not be amenable to surgical flap repair. Superficial radiotherapy is again proving to be an effective means of treatment for non-melanoma skin cancers with a high cure and low recurrence rate. An improved cosmetic outcome has also been noted with minimal hypopigmentation and telangiectasia formation after proper healing.

Unlike larger machines that require a designated treatment room, the superficial radiotherapy machine can be mobilized for use in several areas throughout a clinic. The x-ray tube itself is composed of metal ceramic with several applicator sizes to adjust for the size of the lesion and its margins.

Proton therapy

One of the newest success stories in medical innovation is the proton accelerator. Protons are relatively large subatomic particles that do not scatter as freely as x-rays and electrons. Their energy can be focused on a single point with little dispersion along the way. The figure below shows how the dose is small and constant along the depth of the tissue until it hits a certain predetermined target.

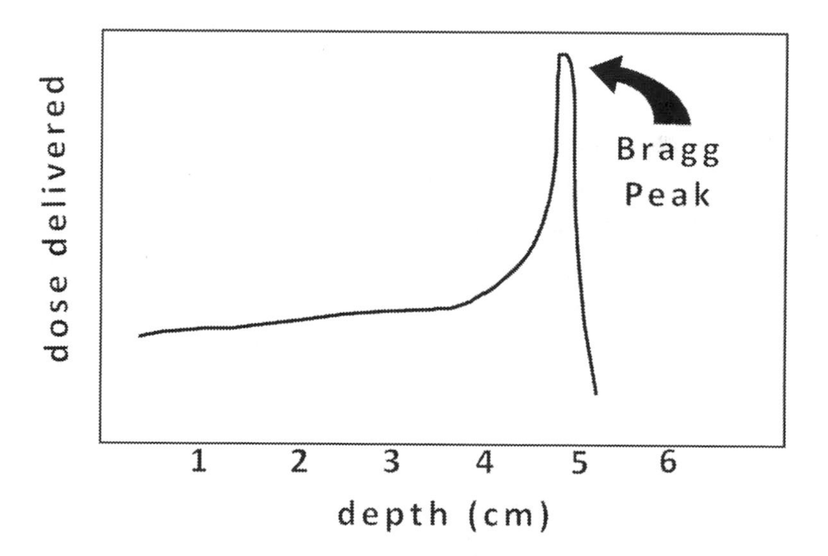

The proton beam's energy is released at a specified depth known as the *Bragg Peak*. This depth is determined by the energy given to the protons when they are initially accelerated. With greater energy comes a deeper Bragg Peak. By adding a modulator, the

Bragg Peak can be spread out to fit the width of the planned treatment volume as shown in the figure below.

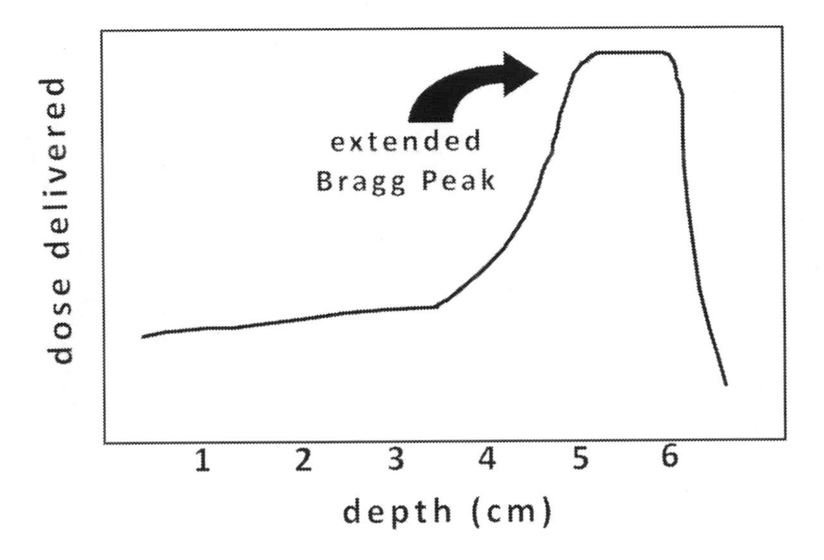

When compared to x-rays, a proton beam does not administer as much radiation to the surrounding tissue as it travels from its point of entry at the skin to its tumor depth. Even more impressive is that the dose delivered beyond the depth of the Bragg peak is very small. In proton therapy, a dose is able to enter the body, act at a certain depth, and virtually disappear. In contrast, an x-ray continues to follow along an exit path, further damaging healthy tissue. Proton therapy is able to focus on the planned treatment volume with little “entrance dose” or “exit dose”. This allows much less damage to critical structures and has made proton therapy a useful tool in prostate, head and neck, brain, and spinal cord tumors.

One example is the treatment of ocular melanoma which traditionally involved enucleation (surgical removal of the eye).

However, proton therapy has been used to improve quality of life and cosmetic appearance in these patients. In addition, a significant decrease in impotence and incontinence is seen among men treated for prostate cancer by proton beam therapy.

The pediatric population is also a focus for proton treatment because traditional radiotherapy causes damage to immature bone and soft tissues. This can result in deformities and growth retardation. With sparing of normal tissue, organ development is less affected and long term side effects of treatment are minimized.

Electron beam therapy (EBT)

Using a linac, electrons can be aimed at superficial tumors in a way similar to protons. There is little damage to underlying structures as electrons only penetrate a small depth, similar to orthovoltage therapy. Many of the same advantages of orthovoltage also apply to EBT. In addition, total body EBT is used to treat cutaneous T-cell lymphoma. This involves treating the entire surface of the body with high energy electrons.

Hyperthermia

Hyperthermia complements radiotherapy by creating vasodilation and, subsequently, better oxygenation to improve cell kill. Research has demonstrated that heating to 40° C for 1 hour can lead to better oxygenation of tumor tissue for up to 24 hours.

This not only promotes radiation cell kill, but also the perfusion of chemotherapeutic agents into tumor microvessels that were otherwise previously inaccessible. In addition, cells are the most radioresistant in the S phase. However, they are also the most sensitive to cell kill by hyperthermia in this phase. The greatest effect of cell kill occurs when radiotherapy is used as soon as possible after hyperthermia.

Hyperthermia works by denaturing proteins, creating chromosomal aberrations. This aids in destroying the DNA repair process. The presence of *heat shock proteins* (proteins that protect the conformational shape of biological molecules from denaturation) protect cells from thermal damage, giving rise to hyperthermic resistance. Adaptation to thermal stress is said to improve when treatment fractions are spaced 48 to 72 hours apart.

Acidification sensitizes cells to hyperthermia therapy, but it has proven difficult to create these conditions within the tumor. Promoting hyperglycemia and administering medications that arrest cellular respiration create acidification. Preliminary research with the use of these agents is promising.

The hyperthermia treatments are accomplished by administering microwaves, radio waves, ultrasound, or other high energy sources via an external source or internally via a probe or needle. Radiofrequency ablation is the most common vehicle used for hyperthermia treatment. It emits high energy radio waves via a thin probe that is placed into the tumor. Probe placement is guided by ultrasound, fluoroscopy, or CT scan. Treatment times are usually about 10 to 15 minutes.

Other clinical uses for radiotherapy

Intraoperative electron beam therapy

Use of radiation during surgical oncology procedures is known as intraoperative electron beam therapy (IOEBT). In the operating room the surgeon removes as much tumor bulk as possible. Inaccessible or microscopic tumor beyond the surgical margins may remain despite the surgeon's best effort. The radiation oncologist then uses electron therapy (given their superficial penetration) to treat. A single dose is delivered to the open patient on the surgical table. This minimizes the dose that is absorbed by normal surrounding tissue. IOEBT can be effectively used in bone and soft tissue sarcomas, gynecologic, bronchogenic, and gastrointestinal cancers.

Palliative radiotherapy

When conservative pain management regimens fail and surgical resection is not feasible, palliative radiotherapy may be initiated for relief. Often used for bone metastases or nerve compression, radiation delivered in accelerated fractionation schedules has been shown to completely ameliorate pain in over half of patients treated. Over 90% report improvement in their condition. Relief is not immediate, but requires approximately 2 weeks before symptoms begin to improve. Regimens may range from a single dose of 800 Gy up to 4000 Gy delivered over 10 fractions. Metastatic bone pain may flare up after palliative radiotherapy, and patients receiving a single fraction are at higher risk of developing flares.

Radiotherapy emergencies

Radiation therapy can be used in life threatening conditions such as spinal cord compression, airway obstruction, and brain tumors at risk for herniation. However, it should be noted that this will require time to take effect, and is often used concomitantly with other therapies.

Notes

Notes

Notes

Notes

Notes

Notes

Notes

Notes

Notes

Notes

Acknowledgments

Jonathan Beitler, Anna & Brent Harris, John & Mallory Stewart, Bryce Gartland, Ranjan Perera, Steve Thombley, Darshan Acharya, Leonard Gyebi, David Krakow, George Mathew, Ira Marsh, Nabil Saba, Taofeek Owonikoko, Jeff Taylor, Than Win, Steve Carpenter, Sidney Smith, Larry Appel, Ellen, Ricky, & Madge Quarles, Bill, Ruby & Deanna Purdie, John McGuirk, & Jill.

Notes

Notes

Acknowledgments

Jonathan Beitler, Anna & Brent Harris, John & Mallory Stewart, Bryce Gartland, Ranjan Perera, Steve Thombley, Darshan Acharya, Leonard Gyebi, David Krakow, George Mathew, Ira Marsh, Nabil Saba, Taofeek Owonikoko, Jeff Taylor, Than Win, Steve Carpenter, Sidney Smith, Larry Appel, Ellen, Ricky, & Madge Quarles, Bill, Ruby & Deanna Purdie, John McGuirk, & Jill.